Design and Stability of Concrete Structures

Structural Engineering

originally prepared
under the direction of
John R. McMahon

C. Brown
Publication Editor

Wexford
2008

Note that this book is based on an earlier publication by the Corps of Engineers, and references other US Army and Corps of Engineers documents, detailed in the back of this book under Appendix A.

Engineering and Design
STABILITY ANALYSIS OF CONCRETE STRUCTURES

Table of Contents

Chapter 4
Loads and Loading Conditions

Chapter 5
Soil Forces and Single Wedge Sliding Analysis

Chapter 6
Stability Considerations and Analytical Methods

Chapter 7
Evaluating and Improving Stability of Existing Structures

Chapter 8
Anchoring Structures

Chapter 1
General

1-1. Purpose

This manual estabilshes and standardizes stability criteria for use in the design and evaluation of the many various types of concrete structures common to both commercial and civil works projects. As used in this manual, the term "stability" applies to external global stability (sliding, rotation, flotation, and bearing), not to internal stability failures such as sliding or lift surfaces or exceedance of allowable material strengths.

1-2. Applicability

This manual applies to both commercial and civil engineering projects, including those of the US Corps of Engineers.

1-3. References

Required and related publications are listed in Appendix A.

1-4. Mandatory Requirements

Designers performing stability analyses of concrete structures are required to satisfy specific mandatory requirements. The purpose of mandatory requirements is to assure the structure meets minimum safety and performance objectives. Mandatory requirements usually pertain to critical elements of the safety analysis such as loads, load combinations and factors of safety. Mandatory requirements pertaining to the guidance contained in a particular chapter are summarized at the end of that chapter. No mandatory requirements are identified in the appendices. Instead, any mandatory requirements pertaining to information contained in appendices is cited in chapters which refer to those appendices. Where other Corps guidance documents are referenced, the designer must review each document to determine which of its mandatory requirements are applicable to the stability analysis. Engineers performing the independent technical review must ensure that the designers have satisfied all mandatory requirements. Waiver procedures for mandatory requirements are described in ER 1110-2-1150. This reference also indicates that deviation from non-mandatory provisions should be rare, and are subject to approval by the engineering chief in the design district.

1-5. Scope

This manual covers requirements for static methods used in stability analyses of hydraulic structures. The types of concrete structures addressed in this manual include dams, locks, retaining walls, inland floodwalls, coastal floodwalls, spillways, outlet works, hydroelectric power plants, pumping plants, and U-channels. The structures may be founded on rock or soil and have either flat or sloped bases. Pile-founded structures, sheet-pile structures, and footings for buildings are not included. When the stability requirements of this manual conflict with those in other Engineering Manuals or Engineering Technical Letters, the requirements of this manual shall govern. These requirements apply to all potential failure planes at or slightly below the structure/foundation interface. They also apply to certain potential failure planes within unreinforced concrete gravity structures. This manual defines the types and combination of applied loads, including uplift forces due to hydrostatic pressures in the foundation material. The manual defines the various components that enable the structure to resist movement, including anchors to the foundation. Most importantly, the manual prescribes the safety factors, which govern stability requirements for the structure for various load combinations. Also, guidance is provided for evaluating and improving the stability of existing structures.

1-6. Background

a. General. Engineer Manuals published over the past 40 years have set stability requirements for the different major civil works structures. For sliding and bearing, the stability requirements have been expressed deterministically in terms of an explicit factor of safety that sets the minimum acceptable ratio of foundation strength along the most critical failure plane to the design loads applied to the failure plane. The analysis for determination of the resultant location in prior guidance has been termed an *overturning stability analysis.* This is a misnomer since a foundation bearing, crushing of the structure toe, and/or a sliding failure will occur before the structure overturns. This manual replaces the term overturning stability analysis with resultant location.

b. Intent. The basic intent of the new guidance specified herein is summarized below:

- Provide new standard factors of safety as replacement for the somewhat variable factors of safety previously specified in other Corps guidance documents.

- Establish basic structural performance goals for each loading condition category.

- Provide tabular summaries of the structure-specific loading-condition check lists found in the other Corps guidance documents in order to properly categorize each loading condition as either usual, unusual, or extreme.

- Require the use of higher factors of safety for conditions where site information is not sufficient to provide a high degree of confidence with respect to the reliability of foundation strength parameters, loads information, and analytical procedures used in the stability analysis.

- Permit the use of lower factors of safety for existing structures when there is a high degree of confidence, based on records of construction and in-service conditions, that the values of the critical parameters used in the stability analysis are accurate.

The process used to standardize factors of safety is based on the premise that the traditional factors of safety specified in the recent guidance for Corps concrete hydraulic structures, for the most part, provide adequate protection against stability failure. The standardization process recognizes, as did previous Corps guidance, that lower factors of safety can be assigned to those loads and loading conditions designated as *unusual,* or *extreme* because the probabilities of those loads and load conditions occurring during the life of the structure are significantly less than the probabilities for *usual* loading conditions. The following elements were part of the safety factor standardization process:

- Traditional factors of safety specified in previous Corps guidance documents were used as a basis for establishing new factors of safety, which are re-formatted to be consistent with other Corps guidance that has probabilistic based requirements.

- The guidance incorporates past practices of assigning lower factors of safety to *normal* structures, as compared to those traditionally used for *critical* structures,.

- The guidance incorporates past practices of categorizing maintenance and construction loads as *unusual* loads.

- The guidance defines the loading condition categories of *usual, unusual,* and *extreme* in probabilistic terms to provide standardization as to which category various structure specific loadings should be assigned.

- Provides a consistent set of safety factors, which account for loading probability, critical structures, and the knowledge of site information used in the stability analysis.

c. Factors of safety. Factors of safety are needed in stability and structural analyses because of the potential variability in loads and material strengths. The factor of safety assigned to a particular stability design or investigation reduces the risk of unsatisfactory performance due to loads being greater than assumed for design and the risk of unsatisfactory performance due to material strengths being less than assumed for design. This guidance makes no attempt to quantify the reliability of the safety factors prescribed for use in the design and evaluation of Corps structures other than that they are traditionally accepted values that when used with prescribed simple assessment procedures have produced structures which have performed satisfactorily for many years. The minimum-allowable safety factors described in this manual assume that a complete and comprehensive geotechnical investigation has been performed. Higher safety factors are required when site information is limited. When concerns about stability exist, the designer should take all measures necessary to quantify load and material strength variability and use the most comprehensive analytical tools available to evaluate the capacity of the structure to meet performance objectives.

d. Sliding stability. Sliding of a structure on its foundation represents the most difficult aspect of a stability analysis, especially in those instances where the foundation is jointed, and where the strength properties vary throughout the foundation. The approach to evaluating sliding stability is one that uses the limit equilibrium method with the linear Mohr-Coulomb failure criterion as a basis for estimating maximum available shear strength. The greatest uncertainties in the analysis are those associated with shear strength determination. The safety factors used are consistent with current foundations and explorations procedures, and with current analytical methods. The guidance recognizes that there are foundations where design shear strength parameters are highly variable because foundation conditions change from one area of the foundation to another and because the foundation may be comprised of both intact rock, and jointed rock with clean or filled discontinuities all with differing shear / displacement characteristics and possibly with strain-softening characteristics which make overall strength a function of displacement. A combination of experience and judgment is necessary to confidently determine that the strength and load parameters used in the stability analysis will provide structures that meet performance objectives.

1-7. Coordination

Even though stability analysis of concrete structures is a structural engineering responsibility, the analysis must be performed with input from other disciplines. It is necessary to determine hydrostatic loads consistent with water levels determined by hydraulic and hydrological engineers. Geotechnical engineers and geologists must provide information on properties of foundation materials, and must use experience and judgement to predict behavior of complex foundation conditions. To ensure that the proper information is supplied, it is important that those supplying the information understand how it will be used by the structural engineer. To ensure that the information is applied appropriately, it is important that the structural engineer understand methods and assumptions used to develop this interdisciplinary data.

Chapter 2
Failure Modes and Wedge Sliding Analysis

2-1. General

The objective of a stability analysis is to maintain horizontal, vertical, and rotational equilibrium of the structure. Geotechnical information is needed to properly define and perform a realistic stability analysis. Possible failure modes and planes of weakness must be determined from onsite conditions, material strengths, and uplift forces. Stability is ensured by:

- Providing an adequate factor of safety against sliding at all possible failure planes.
- Providing specific limitations on the magnitude of the foundation bearing pressure.
- Providing constraints on the permissible location of the resultant force on any plane.
- Providing an adequate factor of safety against flotation of the structure.

However, satisfying the above provisions may not ensure stability if the structure experiences significant loss of foundation material due to erosion or piping, or if there is an internal failure due to inadequate strength of the structural materials. Stability is just one of the requirements necessary to ensure adequate structural performance.

2-2. Limit Equilibrium Analysis

The forces and pressures acting on a structure are indeterminate. Static equilibrium equations are insufficient to obtain a solution for lateral soil forces; additional assumptions must be incorporated in the analysis. For nonlinear materials, such as soils, this is commonly done by assuming that a limit or failure state exists along some surface and that the shear force along the surface corresponds to the shear strength of the material. With these assumptions, equilibrium equations can be solved. Hence, this approach is commonly called limit-equilibrium analysis. To ensure that the assumed failure does not occur, a reduction factor (safety factor or strength mobilization factor) is applied to the material strength. It should be noted that this approach differs significantly from that commonly used for indeterminate structure analysis, where stress-strain properties and deformations are employed. This limit equilibrium approach provides no direct information regarding deformations; it is implied that deformations are sufficient to induce the failure condition. Actual deformations will vary non-linearly in response to actual applied loads. Deformations are indirectly limited to tolerable values by the judicious selection of a safety factor.

2-3. Sliding Planes

Stability must be assessed on selected surfaces within structure in accordance with the methods presented in EM 1110-2-2200. Sliding safety must also be assessed at/or near the foundation-structure interface. This surface may be either level or sloping. Generally, it may be assumed that a surface that slopes upward (in the direction of possible sliding) will have a beneficial effect, while one that slopes downward will increase the possibility for sliding. Figure 2-1 illustrates the beneficial and adverse effects of base slope. Where a shallow weak seam exists below a structure's contact with the foundation, or a structure is imbedded below the top of the foundation, two possible failure modes are present. One mode involves slippage along the weak plane (directly under the structure) and along its extension until it daylights. The other mode involves slippage along the weak plane directly under the structure plus slippage along a plane through the foundation above the weak seam (crossbed shear for rock or passive resistance for soil). When the weak seam extends a large distance past the toe of the structure without daylighting, the second mode will usually be critical. Figure 2-2 illustrates these modes of failure.

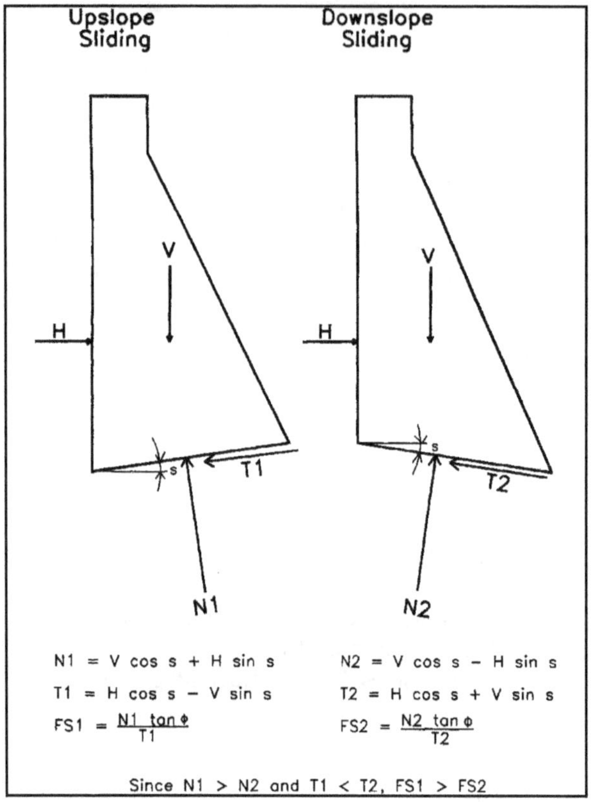

Figure 2-1

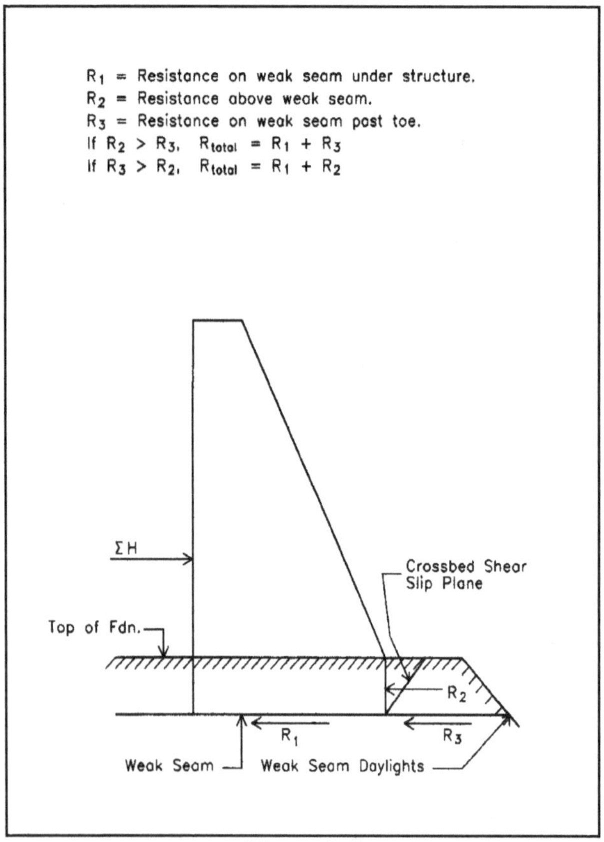

Figure 2-2

2-4. Resultant Location

Conformance with resultant location requirements ensures that the structure is safe from rotational failure. The slope of the resultant and its location are critical in assessing the foundation bearing capacity. For some load condition categories, the resultant is allowed to fall outside the middle-third of the base. In these instances, it is assumed that the structure-foundation interface has no capability for resisting tensile stresses; therefore, part of the structure's base is assumed to lose contact with the foundation resulting in changes to the uplift pressure acting on the base.

2-5. Flotation

This mode of failure occurs when the net uplift force (gross uplift on the base minus the weight of surcharge water above the structure) exceeds the summation of forces due to the weight of the structure, the weight of water contained in the structure, and other surcharge loads.

2.6. Bearing

Analytical methods, traditional bearing capacity equations, field tests, and laboratory tests are all used to determine the bearing capacity of soil and rock. The allowable bearing capacity is defined as the maximum pressure that can be permitted on a foundation soil or rock mass giving consideration to all pertinent factors, with adequate safety against rupture of the soil or rock mass, or settlement of the foundation of such magnitude as to jeopardize the performance and safety of the structure. Increases in allowable bearing capacity are permitted for unusual and extreme load conditions over those required for usual load conditions. The allowable increases are provided in Chapter 3. Shear strength parameters used in the determination of bearing capacity values shall be established in accordance with the discussion presented in Paragraph 2-8.

a. *Soil.* For structures founded on soil, the bearing capacity is limited by the ability of the soil to safely carry the pressure placed on the soil from the structure without undergoing a shear failure. Prevention of a shear failure, however, does not ensure that settlements will be within acceptable limits, therefore, a settlement analysis is usually performed in addition to the shear analysis. Discussion on methods for estimating settlements and limitations in accuracy of settlement analyses is contained in EM 1110-1-1904. Methods for determining allowable bearing capacity of soils are covered in EM 1110-1-1905. General shear failure in a homogeneous soil foundation, for a vertical loading applied at the middle of a structure with a horizontal base-foundation contact, is illustrated by Prandtl's arc of shear failure as shown in Figure 2-3. The shape of the failure surface will be affected by eccentricity, the presence of shear components, and slope of the contact surface. Eccentricity and the presence of shear components will tend to make this type of failure more probable, while a sloping contact surface can either increase or decrease the probability of failure depending upon the slope direction.

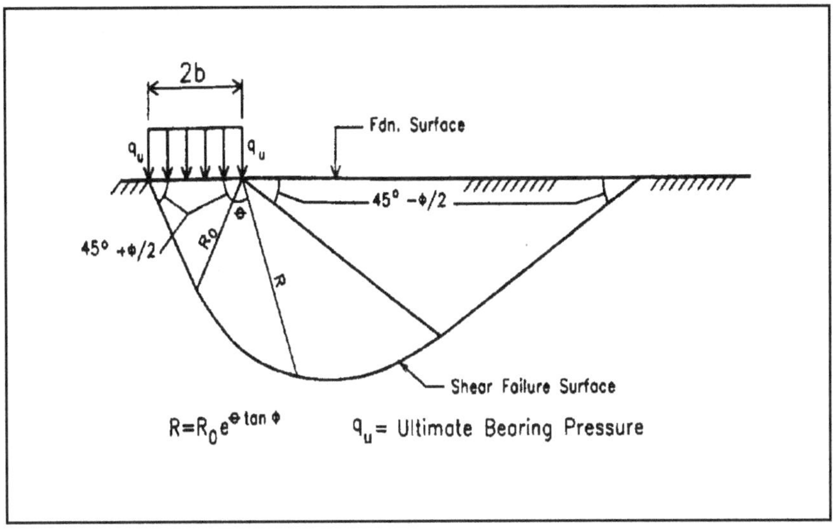

Figure 2-3

b. *Rock.* For structures founded on rock, failure modes may consist of local crushing, shear failures on weak seams, and failures at discontinuities or along bedding planes. The bearing capacity of rock will depend on whether the rock is intact, jointed, layered, or fractured. Methods used for determination of rock bearing capacities are contained in EM 1110-1-2908. The design for structures on rock foundations will involve sliding stability analyses as well as bearing capacity and settlement analyses. Sliding stability analyses address the ability of the rock foundation to resist the imposed loads without the occurrence of shearing on any horizontal or sloping weak plane. Basic rock foundation data that should be obtained for use during the design stage include material properties, strike, dip, thickness, and discontinuities such as faults, fissures and fractures. Such information should be incorporated into the bearing capacity, settlement, and sliding stability analyses.

c. *Coordination between disciplines.* The structural engineer and geotechnical engineer/geologist must coordinate their efforts in order to properly evaluate the bearing capacity of a foundation. Bearing capacity is affected by the size and shape of structure's base, the type of structure, type of loading (static or dynamic), load duration, the eccentricity of the load acting on the foundation, and the shear components of the load; all of which should be furnished to the geotechnical engineer/geologist by the structural engineer. The location and identification of weak zones and planes or discontinuities, soil and rock strength parameters, information on existing faults, and the allowable bearing capacity of the foundation should be furnished to the structural engineer by the geotechnical engineer/geologist.

2-7. Geotechnical Explorations and Testing

The scope of any geotechnical investigation will depend on geological structural complexity, imposed or existing loads acting on the foundation, and to some extent the consequences should a failure occur. The complexity of the foundation will determine the extent of drill holes, mapping, trenching, and other exploratory measures, which may be required to accurately describe foundation conditions. Guidelines for foundation explorations and testing are provided in EM 1110-1-1802, EM 1110-1-1804, and EM 1110-1-2908.

2-8. Shear Strength Tests

Shear strength parameters required for bearing capacity and sliding stability analyses may be estimated for soils from the results of in situ tests and/or by direct shear and triaxial tests performed in the laboratory. For rock, these values

are usually obtained from laboratory tests. Shear strength is a function of many complex independent variables including mineralogy; particle size, shape and gradation; cementation; degree of consolidation; state of stress; anisotropy; and drainage conditions. Therefore, any tests performed in the laboratory should model the conditions that will occur during project operation. Since shearing may take place on any plane that includes intact rock, sheared rock, or jointed rock, strength values for all differing rock conditions must be established in order that sliding stability and bearing capacity may be determined. EM 1110-2-1906 provides guidance on laboratory soils testing. Procedures for testing soils are also described in EM 1110-1-1804 and EM 1110-2-1913. Procedures for testing rock specimens are given in EM 1110-1-2908 and the "Rock Testing Handbook" (U. S. Army Waterways Experiment Station 1980). Coordination must be maintained between the structural engineer and the geotechnical engineer/geologist to ensure that safe and economical designs are obtained.

a. Soils Tests.

(1) *In-situ soils tests.*

- Standard Penetration Test (SPT). The SPT resistance, often referred to as the blow count, is frequently used to estimate the relative density of soil. This relative density can then be correlated with the angle of internal friction, ϕ, and the undrained shear strength, c.

- Cone Penetration Test (CPT). The CPT may also be used to estimate the relative density of cohesionless soils and the undrained shear strength of cohesive soils. The CPT is especially suitable for sands, where it is preferable to the SPT.

- Field Vane Test (FVT). The FVT is commonly used to estimate the in situ undrained cohesive strength of soft to firm clays.

(2) Laboratory soils tests.

- Q-Test. In a Q test the water content of the soil sample is not permitted to change either prior to or during load application. The Q test produces results that approximate the shear strength available for short-term loading conditions. In cohesive soils this test yields relatively large c (cohesion) values and very low or zero ϕ values.

- R-Test. The R test represents conditions in which impervious or semi-pervious soils that have been consolidated under one set of stresses are subjected to a stress change without time for further change in water content prior to failure. In cohesive soil this test furnishes undrained shear strength parameters.

- S-Test. An S test is used to measure drained or effective stress strength parameters, c' and ϕ'. The soil sample is consolidated under an initial confining stress and loading increments are applied slowly enough to allow pore water pressures to dissipate with each load increment during shear. Results of S tests are applicable to free-draining soils where excess pore pressures do not develop during shear. S tests are also used for evaluating the shear strength of cohesive soils under long term loading conditions, where excess pore pressures have dissipated.

b. Rock Tests. The tests used for evaluating rock shear strength parameters should mirror, as closely as possible, the conditions that are expected to exist in the field. The structural engineer is referred to EM 1110-1-2908, and numerous references contained therein, in order to become familiar with such things as rock descriptors, rock mass classification systems, laboratory classification and index tests for rock, selection of modulus of deformation, etc.

(1) In-situ rock tests.

- In-Situ Direct Shear Test. In-situ direct shear tests are expensive and are only performed where critically located, thin, weak, continuous seams exist within relatively strong adjacent rock. Relatively large surface areas must be tested in order to address unknown scale effects. The test, as performed on thin, fine grained, clay seams, is considered to be an undrained test.

- In-Situ Uniaxial Compression Test. In-situ uniaxial compressive tests are expensive. This test is used to measure the elastic properties and compressive strength of large volumes of virtually intact rock in an unconfined state. The results obtained are useful in evaluating the effects of scale, however, the test is seldom used just for this purpose.

(2) Laboratory rock tests.

- Unconfined Uniaxial Compression Test. This test is performed primarily to obtain the unconfined compression strength and the elastic properties of a rock sample. Poisson's ratio can be determined if longitudinal and lateral strain measurements are taken during the test. Occasionally samples are tested with differing orientations in order to describe three-dimensional anisotropy. This test may not be indicative of the overall rock mass strength.

- Triaxial Test. The triaxial test can be made on intact, cylindrical rock samples. Data is provided for determination of rock strength in the drained or undrained state when subjected to three-dimensional loading. Data from the test, used in calculations, provides the strength and elastic properties of the sample at various confining pressures. Strengths along planes of weakness, such as natural joints, seams, and bedding, can be determined if these planes are properly oriented in the test. The oriented plane variation is particularly useful for strengths on thinly filled discontinuities containing soft materials. Since clean discontinuities are free draining, tests performed on them should be drained tests. Tests on discontinuities filled with coarse-grained materials should also be drained tests. The tests for discontinuities filled with fine-grained materials should be undrained tests.

- Laboratory Direct Shear Test. The laboratory direct shear test is primarily used to measure the shear strength, at various normal stresses, along planes of discontinuity or weakness. When this test is performed on the surface of a clean discontinuity (with asperities) that is subjected to very high normal stress, with a rapidly applied shear stress and small deformations, the values obtained will represent the *peak shear strength*. The test is often performed at a reduced rate of shear stress application, with intermediate or low normal stresses, and with the asperities over-ridden, resulting in reduced values for shear strength. Repetitive shearing of a sample, or continuing displacement to a point where shear strength is constant, establishes the *residual shear strength* that is available. For the test results to be valid, test conditions must be as close as possible to the conditions that will exist in the field. Test drainage conditions should be essentially the same as for the triaxial tests discussed above. *Upper bound* and *lower bound* shear strengths are discussed in EM 1110-1-2908.

2-9. Selection of Design Shear Strengths

Design shear strength parameters should be selected by the geotechnical engineer/geologist in consultation with the structural engineer. All parties must be aware of the implicit assumptions pertaining to the stability analysis procedure that is being used. The design strengths should be selected such that they do not result in uneconomical, ultra-conservative designs. However, in certain instances it may be more economical to assume conservative design shear strength parameters than to institute an expensive testing program. Selected strength parameters must be appropriate for the actual stress states and drainage conditions expected for the foundation materials. Laboratory results are dependent on the details of the testing procedures and the condition of the samples tested. The conversion from laboratory test data to in-situ strength parameters requires careful evaluation. A combination of experience and judgment is required to give the level of confidence needed for selecting the strength parameters. The shear strength of intact rock as well as rock with clean or filled discontinuities is dependent on many factors including confining pressures, loading history, and rate of loading. Specimen size is also a factor which must be considered when estimating shear strength based on laboratory testing. The number, orientation, and size of discontinuities and weaknesses may vary considerably, thus affecting load distribution and the final results. When selecting design shear strengths, the shape of stress-strain curves for individual tests should be considered. Where undisturbed and compacted samples do not show a significant drop in shear or deviator stress after the peak stress is reached, the design strength can be chosen as the peak shear stress. Where significant differences in stress-strain characteristics exist along a potential failure surface, stress-strain compatibility and the potential for progressive failure must be considered in selection of design strength parameters. With varied foundation conditions, it may not be possible to have all the foundation materials at their peak strengths at the same displacement (see Chapter 6). In those conditions, and for conditions that rely on passive resistance of a rock wedge or soil backfill, the strength values must be consistent with the displacements that will put the structure at the limit state assumed for the sliding stability analysis. A discussion of the sliding equilibrium method and its limitations can be found in EM 1110-1-2908.

2-10. Multiple-Wedge Sliding Analysis

a. Basic concepts. The multiple-wedge sliding analysis is a fairly simple assessment of the sliding factor of safety along various potential sliding planes, that can also account for the behavior expected from complex soil stratification and geometry. It is based on modern principles of structural and geotechnical mechanics that apply a safety factor to the material strength parameters in a manner, which places the forces acting on the structure and foundation wedges in sliding equilibrium. This method of analysis is illustrated in Appendix D, example D2. Derivation of the governing equilibrium equation for a typical wedge is shown in Appendix E. See EM 1110-1-2908 for additional information on this method. Following are the principles, assumptions and simplifications used in multiple-wedge sliding analysis.

- Sliding stability of most concrete structures can be adequately assessed by using a limit equilibrium approach.

- A sliding mode of failure will occur along a presumed failure surface when the applied shearing force exceeds the resisting shearing forces.

- The failure surface can be any combination of plane and curved surfaces, but for simplicity, all failure surfaces are assumed to be planes, which form the bases of wedges.

- Analyses are based on assumed-plane failure surfaces. The calculated safety factor will be realistic only if the assumed failure mechanism is possible.

- The factor of safety is defined, and minimum required factors of safety are given in Chapter 3.

- The lowest safety factor on a given failure surface can be determined by an iterative process. However, a single-step analysis using the required minimum factor of safety, can be used as a simple pass/fail test.

- A two-dimensional analysis is presented in this manual. These principles should be extended if unique three-dimensional geometric features and loads critically affect the sliding stability of a specific structure.

- Only force equilibrium is satisfied in this analysis, moment equilibrium is not ensured.

- The shearing force acting along the vertical interface between any two wedges is assumed to be negligible. Therefore, the failure surface at the bottom of each wedge is only loaded by the forces directly above it.

- A linear relationship is assumed between the resisting shearing force and the normal force acting on the failure surface beneath each wedge.

- The maximum shear strength that can be mobilized is adequately defined by the Mohr-Coulomb failure theory.

- Considerations regarding displacements are excluded from the limit equilibrium approach. The relative rigidity of different foundation materials and the concrete substructure may influence the results of the sliding-stability analysis. Such complex structure-foundation systems may require a more intensive sliding investigation than a limit equilibrium approach. The effects of strain compatibility along the assumed failure surface may be included by interpreting data from in situ tests, laboratory tests, and finite element analyses.

b. Analytical procedure. Following is a general procedure for analyzing multi-wedge systems.

- Assume a potential failure surface which is based on the stratification, location and orientation, frequency and distribution of discontinuities of the foundation material, and the configuration of the substructure. Discontinuities in the slip path beneath the structural wedge should be modeled by assuming an average slip plane along the base of the structural wedge. The structural wedge may include rock or soil that lies below the base of the concrete structure.

- Divide the assumed slide mass into a number of wedges. Since all portions of the structure must slide as a unit, there can be only a single structural wedge.

- The interface between the group of driving wedges (wedges with negative slip plane inclination angles) and the structural wedge is assumed to be a vertical plane located at the heel of the structural wedge and extending to the base of the structural wedge. The magnitudes of the driving forces depend on the actual values of the safety factor and the inclination angles (α) of the slip path. The inclination angles, corresponding to the maximum driving forces for each potential failure surface, can be determined by independently analyzing the group of driving wedges for a trial safety factor. In rock, the inclination may be predetermined by discontinuities in the foundation. The general equation applies to wedges with lateral earth forces that act horizontally.

- The interface between the group of resisting wedges (wedges with positive slip plane inclination angles) and the structural wedge is assumed to be a vertical plane located at the toe of the structural wedge and extending to the base of the structural wedge. The magnitudes of the resisting forces depend on the actual values of the safety factor and the inclination angles of the slip path. The inclination angles, corresponding to the minimum resisting forces for each potential failure mechanism, can be determined by independently analyzing the group of resisting wedges for a trial safety factor. When resisting force is used, special considerations may be required. Rock that may be subjected to high velocity water scouring should not be used unless adequately protected. Also, the compressive strength of the rock layers must be sufficient to develop the wedge resistance. In some cases, wedge resistance should not be assumed without resorting to special treatment such as installing rock anchors.

- Draw free body diagrams which show all the forces assumed to be acting on each wedge. The orientation of the failure surfaces for most wedges can be calculated directly by using the equations in paragraph 5-4.

- The analysis proceeds by assuming trial values of the safety factor and unknown inclinations of the slip path so the governing equilibrium conditions, failure criterion, and definition of safety factor are satisfied. An analytical or a graphical procedure may be used for this iterative solution.

- If it is only necessary to determine whether an adequate safety factor exists, this may be determined in a single step without the iterative process.

- For some load cases, the normal component of the resultant applied loads will lie outside the kern of the base area, and a portion of the structural wedge will not be in contact with the foundation material. The sliding analysis should be modified for these load cases to reflect the following secondary effects due to coupling of sliding and rotational behavior. The uplift pressure on the portion of the base, which is not in contact with the foundation material, should be a uniform value, which is equal to the hydrostatic pressure at the adjacent face, (except for instantaneous load cases such as due to seismic forces). The cohesive component of the sliding resistance should only include the portion of the base area, which is in contact with the foundation material.

c. Coordination. An adequate assessment of sliding stability must account for the basic structural behavior, the mechanism of transmitting compressive and shearing loads to the foundation, the reaction of the foundation to such loads, and the secondary effects of the foundation behavior on the structure. A fully coordinated team of geotechnical and structural engineers and geologists should ensure that the result of the sliding analyses is properly integrated into the overall design. Some of the critical aspects of the design process which require coordination are:

- Preliminary estimates of geotechnical data, subsurface conditions, and types of structures.

- Selection of loading conditions, loading effects, potential failure mechanisms, and other related features of the analytical models.

- Evaluation of the technical and economic feasibility of alternative structures.

- Refinement of the preliminary substructure configuration and proportions to consistently reflect the results of detailed geotechnical site explorations, laboratory testing, and numerical analyses.

- Modification of the structure configuration or features during construction due to unexpected variations in the foundation conditions.

2-11. Single-Wedge Sliding Analysis

Only the structural wedge is actively considered in the single-wedge sliding analysis. This is a simpler method, which will produce satisfactory results. The basic concepts are similar for both the single and multiple-wedge methods, but all driving and resisting wedges are replaced with earth and groundwater forces calculated directly, using the methods in Chapter 5. The single-wedge method is illustrated in Appendix D, example D1. This method produces reasonably conservative estimates of the earth forces used for the sliding analysis and for other stability analyses and for structural design.

Chapter 3
Stability Requirements

3-1. General

The concepts used to develop the structural stability requirements contained in this manual are to establish safety factors or safety provisions for the three prescribed load condition categories of usual, unusual, and extreme such that the risk of a failure is kept to an acceptably low level and such that performance objectives are achieved. The use of three different design-load condition categories permits different safety factors or safety provisions to be assigned to the various load conditions depending on the probability of the load condition occurring during the life of the structure. The load conditions used in the stability analyses are described on a probabilistic basis, except the seismic loads and large flood loads falling into the extreme category may be either probabilistic or deterministic.

3-2. Load Condition Categories

The load conditions that a structure may encounter during its service life are grouped into the load condition categories of usual, unusual, and extreme. Associated with each category is a likelihood that the load condition will be exceeded in a given time period. The load conditions, expressed in probabilistic terms, are provided in Table 3-1. The structural performance and the risk of damage or failure depends not only on the likelihood of the loading condition, but also on the safety factors or the safety provisions used, the degree of conservatism used in selecting the foundation strength parameters and hydrological data, and the degree of conservatism inherent in the methods used for the analysis. No attempt has been made to define the likelihood of damage or failure in probabilistic terms. However, the use of these guidelines in conjunction with other Corps guidance will provide structures with adequate protection against stability failure.

Table 3-1 Load Condition Probabilities

Load Condition Categories	Annual Probability (p)	Return Period (t_r)
Usual	Greater than or equal to 0.10	Less than or equal to 10 years
Unusual	Less than 0.10 but greater than or equal to 0.0033	Greater than 10 years but less than or equal to 300 years
Extreme	Less than 0.0033	Greater than 300 years

- *Usual* loads refer to loads and load conditions, which are related to the primary function of a structure and can be expected to occur frequently during the service life of the structure. A usual event is a common occurrence and the structure is expected to perform in the linearly elastic range.

- *Unusual* loads refer to operating loads and load conditions that are of infrequent occurrence. Construction and maintenance loads, because risks can be controlled by specifying the sequence or duration of activities, and/or by monitoring performance, are also classified as unusual loads. Loads on temporary structures which are used to facilitate project construction, are also classified as unusual. For an unusual event some minor nonlinear behavior is acceptable, but any necessary repairs are expected to be minor.

- *Extreme* loads refer to events, which are highly improbable and can be regarded as emergency conditions. Such events may be associated with major accidents involving impacts or explosions and natural disasters due to earthquakes or flooding which have a frequency of occurrence that greatly exceeds the economic service life of the structure. Extreme loads may also result from the combination of unusual loading events. The structure is expected to accommodate extreme loads without experiencing a catastrophic failure, although structural damage which partially impairs the operational functions are expected, and major rehabilitation or replacement of the structure might be necessary.

Appendix B lists the loading conditions that must be evaluated to ensure the stability of specific structure types. The loading conditions have been taken from other USACE manuals and may have been modified to be consistent with other provisions of this manual. When a loading condition is defined in terms of a return period (for example, the Operational Basis Earthquake is defined as an earthquake with a return period of 144 years), the structural engineer can determine if the load condition is usual, unusual, or extreme by referring directly to Table 3-1. When a load condition is stated in non-probabilistic terms, (for example, pool elevation at the top of closed spillway gates, or water to the top of a flood wall), the return period must be determined to see if that particular load condition is usual, unusual, or extreme. In some cases, the load condition category is specifically designated based on established practice, irrespective of any return period (for example, construction is listed as an unusual loading). The engineer only needs to verify stability for those conditions listed in Appendix B. For example, for the unusual category, it is not necessary to verify stability for a 300 year flood or earthquake if these are not specifically listed in Appendix B. Definitions of common loadings for civil works projects are provided in Chapter 4, including: normal operating, infrequent flood, maximum design flood, probable maximum flood, operational basis earthquake, maximum design earthquake, and maximum credible earthquake.

3-3. Risk-based Analysis for USACE Flood Project Studies

USACE policy now requires the application of risk-based analysis in the formulation of flood-damage-reduction projects. The requirements are briefly discussed in the next paragraph to familiarize the structural engineer with the procedures used by hydrology/hydraulics (H&H) engineers use to develop the degree of protection provided by the project (i.e., dam height, floodwall height). The structural engineer needs to coordinate with the H&H engineers to obtain return periods for the required loading conditions to determine the load condition category from Table 3-1.

Risk-based analysis quantifies the uncertainty in discharge-frequency, elevation (stage)-discharge, and elevation-damage relationships and explicitly incorporates this information into economic and performance analyses of alternatives. The risk-based analysis is used to formulate the type and size of the optimal structural (or nonstructural) plan that will meet the study objectives. USACE policy requires that this plan be identified in every flood-reduction study it conducts. This plan, referred to as the National Economic Development Plan (NED), is the one that maximizes the net economic benefits of all the alternatives evaluated. It may or may not be the recommended plan, based on additional considerations. A residual risk analysis for the NED Plan is next performed to determine the consequences of exceeding project capacity. For any flood-protection project, it is possible that project capacity may be exceeded sometime during its service life. Therefore, the question becomes, "If that capacity is exceeded, what are the impacts, both in terms of economics and the threat to human life?" If the project-induced and/or residual risk is unacceptable, and a design to reduce the risk cannot be developed, other alternatives are further analyzed. Either a larger project, that will ensure sufficient time for evacuation, or a different type of project, with less residual risk, should be selected to reduce the threat to life and property. For a detailed discussion of the H&H requirements, see ER 1105-2-101 and EM 1110-2-1619.

When the type and size of the project have been selected, detailed design begins. The structural engineer, in coordination with the hydrology/hydraulic engineers, may use expected values (best estimates) of discharge-frequency and stage-discharge curves to estimate return periods for the various prescribed structure-dependent hydrostatic load conditions listed in Appendix B. For load conditions with prescribed water elevations, (for example, water to the top of closed spillway gates, or water to the top of a flood wall) the headwater elevation may be used in conjunction with the stage-discharge curve and discharge-frequency curves to estimate the annual probability and return period for the event representing the load condition. For some projects, such as high pools at power projects, other information such as project operating data will also be used in estimating the return period for a prescribed loading condition. The designer then refers to Table 3-1 to determine if each particular load condition is usual, unusual, or extreme.

3-4. Site Information

a. General. A proper stability analysis cannot be performed without knowing the potential planes of weakness beneath the structure, the strength of the materials along potential planes of weakness, uplift forces that occur on the structure or on planes of weakness, the strength of backfill materials, and all loads and load conditions to which the structure may be subjected. Knowledge of geologic formations beneath the structure is also important in defining seepage conditions and uplift pressures. Without adequate foundation explorations and testing, the safety factors provided to assess stability of the structure are meaningless. Preliminary stability analyses are useful to identify design parameters, which require special attention. In some rock foundations there may be many faults, shear zones, and discontinuities that make it impossible to do little more than predict average shear and cohesive strengths of the materials that make up the foundation. Use of lower bound values for foundation shear strength or upper bound values for loads is only acceptable when it can be demonstrated that the added costs to improve the accuracy of the strength and loading data will not lead to significant savings for the structure or foundation. Lower factors of safety are permitted by this manual in cases where there is an abundance of information on the various foundation and structure properties used to establish the strength parameters for the stability analysis. Conversely, higher factors of safety are required when there is only limited information on either foundation or structure properties. Three categories of site information, *well defined, ordinary*, and *limited,* were used in establishing safety requirements.

b. Well-defined site information. This category is restricted to use for existing projects. To qualify as well defined, site information must satisfy the following requirements:

- Available records of construction, operation, and maintenance indicate the structure has met all performance objectives for the load conditions experienced.
- Foundation strengths can be established with a high level of confidence.
- The governing load conditions can be established with a high level of confidence.
- Uplift pressures for design load conditions are known, or can be extrapolated for design load conditions based on measured uplift pressure data.

c. Ordinary site information. This category applies to most new project designs. To qualify as ordinary, site information must satisfy the following requirements:

- Foundation strengths have been established with current USACE explorations and testing procedures.
- Foundation strengths can be established with a high level of confidence.
- The governing load conditions can be established with a high level of confidence.

d. Limited site information. This category applies to those new or existing structures designated as *normal* (*critical* structures can not be designed or evaluated based on *limited* site information), where either of the following are true:

- Foundation strengths are based on limited or inadequate explorations and testing information, or
- Governing load conditions cannot be established with a high level of confidence because of insufficient historical data on stream flow, flood potential, etc.

3-5. Critical Structures

Civil works structures, for the purpose of establishing safety factors or safety provisions for use in stability analyses, are to be designated as either critical or normal. Structures designated as critical are those structures on high hazard projects whose failure will result in loss of life. Loss of life can result directly, due to flooding or indirectly from secondary effects. Loss of life potential should consider the population at risk, the downstream flood wave depth and velocity, and the probability of fatality of individuals within the affected population. Information is provided in Appendix H to help design engineers determine if the structure should be designated critical or normal.

3-6. Existing Structures

The safety factors provided in this manual are based on the assumption that for critical and normal structures, the strength of the materials in the foundation and structure has been conservatively established through explorations and testing. This may not be the case for older existing structures, or, if adequate explorations and testing were performed, the records may not be available. When the stability of an existing structure is in question, a phased, systematic approach to evaluating stability should be performed before any remedial actions are undertaken to improve stability. This systematic evaluation process is described in Chapter 7. The load conditions used to evaluate an existing structure should be carefully checked to make sure that what was considered as a usual load condition for the original design is not, once the probabilities of the load conditions are examined, really an unusual or extreme load condition. Evaluation of existing structures should utilize analytical methods which accurately describe the behavior without introducing excess conservatism. When available, actual uplift pressures can be used as a basis for evaluating the stability of existing structures.

3-7. Factors of Safety for Sliding

Analysis of sliding stability is discussed in detail in Chapter 2 and Chapter 5. A factor of safety is required in sliding analyses to provide a suitable margin of safety between the loads that can cause instability and the strength of the materials along potential failure planes that can be mobilized to prevent instability. The factor of safety for sliding is defined by equation 3-1. The required factors of safety for sliding stability for *critical* structures and for *normal* structures are presented in Tables 3-2 and 3-3, respectively.

$$FS_s = \frac{N \tan \phi + cL}{T} \qquad (3\text{-}1)$$

where

N = force acting normal to the sliding failure plane under the structural wedge.
ϕ = angle of internal friction of the foundation material under the structural wedge.
c = cohesive strength of the foundation material under the structural wedge.
L = length of the structural wedge in contact with the foundation.
T = shear force acting parallel to the base of the structural wedge.

Table 3-2 Required Factors of Safety for Sliding - Critical Structures

	Load Condition Categories		
Site Information Category	**Usual**	**Unusual**	**Extreme**
Well Defined	1.7	1.3	1.1
Ordinary	2.0	1.5*	1.1*
Limited**	-	-	-

*For preliminary seismic analysis without detailed site-specific ground motion, use FS=1.7 for unusual and FS=1.3 for extreme. See further explanation in section 3.11 b.
**Limited site information is not permitted for critical structures

Table 3-3 Required Factors of Safety for Sliding - Normal Structures

Site Information Category	*Load Condition Categories* Usual	Unusual	Extreme
Well Defined	1.4	1.2	1.1
Ordinary	1.5	1.3	1.1
Limited	3.0	2.6	2.2

3-8. Factors of Safety for Flotation

A factor of safety is required for flotation to provide a suitable margin of safety between the loads that can cause instability and the weights of materials that resist flotation. The flotation factor of safety is defined by equation 3-2. The required factors of safety for *flotation* are presented in Table 3-4. These flotation safety factors apply to both *normal* and *critical* structures and for all site information categories.

$$FS_f = \frac{W_S + W_C + S}{U - W_G} \tag{3-2}$$

where

W_S = weight of the structure, including weights of the fixed equipment and soil above the top surface of the structure. The moist or saturated unit weight should be used for soil above the groundwater table and the submerged unit weight should be used for soil below the groundwater table.
W_C = weight of the water contained within the structure
S = surcharge loads
U = uplift forces acting on the base of the structure
W_G = weight of water above top surface of the structure.

Table 3-4 Required Factors of Safety for Flotation – All Structures

Site Information Category	*Load Condition Categories* Usual	Unusual	Extreme
All Categories	1.3	1.2	1.1

3-9. Limits on Resultant Location

The factor of safety approach established for sliding and flotation is not appropriate for use in the evaluation of rotational modes of failure. Rotational behavior is evaluated by determining the location of the resultant of all applied forces with respect to the potential failure plane. This location can be determined through static analysis. Limits on the location of the resultant are provided in Table 3-5. The entire base must be in compression for the usual load condition, to maintain full contact between the structure and the foundation, so there is no chance for higher uplift pressures to develop in a crack. This helps ensure linear behavior for common loading conditions. For the unusual load case, higher uplift pressures may develop in a relatively short crack, but this would cause only minor nonlinear behavior. For extreme load conditions on typical civil works projects, a shear or bearing failure will occur before overturning could occur. Therefore, the resultant is permitted to be anywhere within the base, and safety is ensured by the safety factor requirements for sliding and by the limits on allowable bearing stresses.

Table 3-5 Requirements for Location of the Resultant – All Structures

	Load Condition Categories		
Site Information Category	**Usual**	**Unusual**	**Extreme**
All Categories	100% of Base in Compression	75% of Base in Compression	Resultant Within Base

3-10. Allowable Bearing Capacity

Allowable concrete compressive stresses and/or allowable bearing capacity values established by materials engineers and geotechnical engineers are used as the basis for evaluating bearing modes of failure. The allowable bearing capacity value is defined as the maximum pressure that can be permitted on soil or rock giving consideration to all pertinent factors with adequate safety against rupture of the soil or rock mass, or movement of the foundation of such magnitude that the structure is impaired. Bearing failure is related to the relative compressibility of the foundation materials, the loading conditions, the geometry of the structure base, and the strength of the foundation and concrete at the structure-foundation interface. Bearing capacity may be related to the shear capacity of the foundation materials or to the deformability of the foundation. Information on foundation bearing analysis can be found in EM 1110-1-1905 for soils, and EM 1110-1-2908 for rock. Safety against bearing failure is generally expressed in terms of an allowable compressive stress for concrete and an allowable bearing capacity for foundation materials. These allowables include an adjustment, which represents a factor of safety. The allowable compressive stress and allowable bearing capacity values are established by testing performed by materials engineers and geotechnical engineers. Discussion on exploration and testing can be found in Chapter 2. The allowable compressive stress and bearing capacity values established for usual load conditions can be increased for the unusual and extreme load conditions. A 15% increase is permitted for unusual load conditions and a 50% increase is permitted for extreme load conditions.

3-11. Seismic Stability

a. General. Traditionally, the seismic coefficient method has been used to evaluate the stability of structures subjected to earthquake ground motions, but this method fails to take into account the true dynamic characteristics of the structure. There have been cases where structures similar to those used on civil works projects have failed during earthquakes because of a sliding or bearing failure. These failures for the most part are attributable to liquefaction and soil strength degradation in the foundation or backfill materials. Seismic stability analyses should be a performed in phases in accordance with requirements of ER 1110-2-1806. Seismic loads to be used in the first phase analysis are provided in Chapter 5 of this manual. Structures which meet sliding stability factor of safety requirements when evaluated by this procedure are considered to be safe and no additional seismic stability analyses are required. Structures that fail to meet factor of safety requirements when evaluated using this procedure should not be considered unsafe or in need of a stability retrofit. The failure to meet these requirements should only suggest the need for other seismic coefficient and dynamic analyses to fairly assess the demands placed on the structure and foundation during a major earthquake. From these advanced analyses engineers can determine if the displacements and stresses experienced by the structure and foundation will place the structure at risk of a stability failure. In many instances, it is acceptable for sliding and rocking to occur at the base of the structure during extreme earthquake load conditions. Stability in such cases is evaluated using dynamic analysis methods, and performance is ensured by limiting permanent displacements to acceptable levels.

b. Modified Factor of Safety. The factors of safety given in Tables 3-2 include FS=1.5 for unusual and FS = 1.1 for extreme load conditions, for ordinary site information. The ordinary site information and related factor of safety must be used in the seismic coefficient method. These factors of safety are based on use of extreme loads with very low probabilities of being exceeded. When factors of safety for seismic loadings are being calculated using the coefficient method, the MCE loads are usually not based on detailed site-specific seismic data. Since the loads would be based on less precise data, there would be greater probability that the predicted extreme loads could be exceeded, therefore, it is appropriate to use higher factor of safety for such analyses. For such analyses, use a factor of safety of 1.7 for unusual and 1.3 for extreme, as stated in the notes following the above table.

3-12. Mandatory Requirements

For a general discussion on mandatory requirements, see Paragraph 1-5. As stated in that paragraph, certain requirements within this manual are mandatory. The following are mandatory for Chapter 3.

a. Load condition categories. Unless the loading condition category (usual, unusual, extreme) is specifically designated in Appendix B, the return period range limitations specified in Table 3-1 shall be used to establish the correct loading condition designation. When the return period for a particular loading condition can not be established with sufficient accuracy to determine if the loading condition is usual or unusual (or unusual or extreme), the loading condition with the more stringent safety requirements shall be used.

b. Critical structures. In accordance with section 3-5, structures on high hazard projects shall be considered *critical* where failure will result in loss of life; all other structures will be classified as *normal.* In making the determination of critical or normal, the engineer must follow the guidelines in Appendix G.

c. Site information. Structures shall be assigned to one of three site information categories: *well-defined, ordinary*, or *limited.* Site information category selection shall be in accordance with the provisions of Paragraph 3-4.

d. Sliding stability. Sliding stability factors of safety shall be equal to, or greater than, the values specified in Tables 3-2 and 3-3. The sliding stability factor of safety shall be determined using Equation 3-1.

e. Flotation stability. Flotation factors of safety shall be equal to, or greater than, the values specified in Table 3-4. The flotation stability factor of safety shall be determined using Equation 3-2.

f. Resultant location. The location of the resultant of all forces acting on the base of the structure shall be within the limits specified in Table 3-5.

g. Bearing pressures. Bearing pressures for *usual* load conditions shall be within allowable limits established by the geologist/geotechnical engineer. Increases in allowable bearing pressures shall not exceed 15% for *unusual* and 50% for *extreme* load conditions, in accordance with the guidance in section 3-10.

h. Loading conditions. As a minimum, the loading conditions provided in Appendix B shall be satisfied in the stability analysis.

i. Loads. Loads shall comply with the mandatory requirements of Chapters 4 and 5.

Chapter 4
Loads and Loading Conditions

4-1. General

Previously, stability criteria was provided in separate manuals for each type of structure. Those manuals listed all of the load cases (loading conditions), which had to be investigated as part of the stability analysis. Those loading conditions are now summarized in tables provided in Appendix B of this manual. The tables list the loading condition and give a classification as usual, unusual, or extreme, as defined in Table 3-1. Following each table are brief descriptions of the loading conditions. The loading conditions have been revised in some cases for general consistency with the provisions of this manual, especially to comply with current practice for flood and seismic loadings. This chapter defines most of the types of loads that are combined to form each loading condition. However, soil loads are defined in Chapter 2 for multiple wedge sliding analyses and in Chapter 5 for single wedge sliding analyses.

4-2. Construction

Based on past practice, construction loading conditions shall be classified as *unusual*, regardless of duration.

4-3. Water Loading Conditions

a. General. All water loading conditions should be based on hydrologic information, which gives median water elevations in terms of return periods. A typical flood hazard curve is illustrated in Figure 4-1. Curves for both headwater and coincident tailwater will be necessary to determine the water loads for dams and navigation locks. Hydraulic engineers commonly use the 90-percent confidence level hazard curve when determining flood protection requirements. However, for stability analysis, structural engineers require median flood hazard curves, which can also be provided by the hydraulic engineers. Based on the information presented in Figure 4-1 a flood pool elevation equal to 21 meters (68.9 feet) would be used to determine the maximum *unusual* loading.

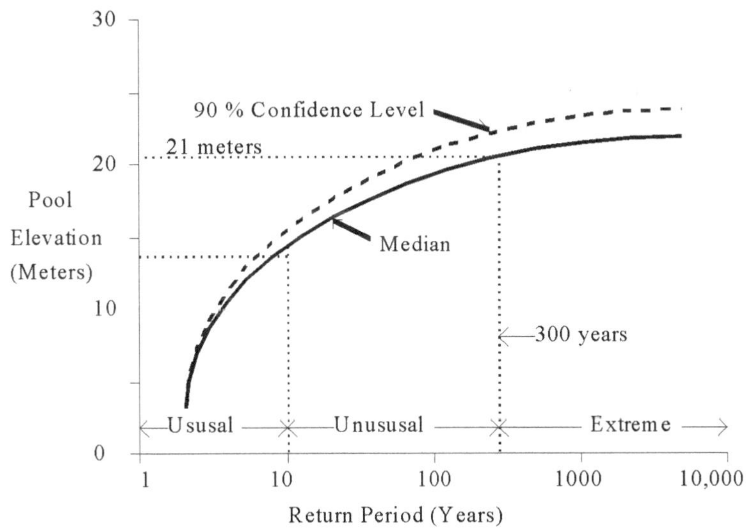

Figure 4-1 Flood Hazard Curve

b. Coincident pool. Coincident pool represents the water elevation that should be used for combination with seismic events. It is the elevation that the water is expected to be at or below for half of the time during each year.

c. Normal operation. In the past, a normal operation loading condition has been used to describe loadings with various probabilities of occurring, including rare events with long return periods. To be consistent with Table 3-1, normal operating conditions are now defined as maximum loading conditions with a return period of no more than 10 years (annual probability of 10%). For certain floodwalls, this means that there might be no water loads on the structure for normal operation. For hydropower dams, the pool will be fairly high for normal operation, while for some flood-control dams, the pool will be low for normal operation. For navigation projects, the maximum loading for normal operation might correspond to the usual navigation pool, combined with the lowest tailwater expected with a 10-year return period. Water loads defined by the normal operation loading condition are sometimes combined with other types of events (such as barge impacts).

d. Infrequent flood. The infrequent flood (IF) represents flood pool or water surface elevations associated with events with a return period of no greater than 300 years (annual probability of 0.33%), making the IF an unusual loading per Table 3-1. This loading condition replaces loadings such as water *to top of spillway gates* and *water to spillway crest* previously used for the design and evaluation of gated and ungated spillways. It also replaces the *design flood* (top of wall less freeboard) used for the design and evaluation of floodwalls. In limited cases, historical hydrologic data may be inadequate to determine the 300-year water elevations with reasonable certainty. In such cases, traditional loading conditions such as *water to top of spillway gates, water to spillway crest,* and *design flood* shall be considered unusual events and evaluated in addition to the IF event.

e. Probable maximum flood. The probable maximum flood (PMF) is one that has flood characteristics of peak discharge, volume, and hydrograph shape that are considered to be the most severe reasonably possible at a particular location, based on relatively comprehensive hydro-meteorological analyses of critical runoff-producing precipitation, snow melt, and hydrologic factors favorable for maximum flood runoff. The PMF load condition represents the most severe hydraulic condition, but because of possible overtopping and tailwater effects, it may not represent the most severe structural loading condition, which is represented by the maximum design flood described below. Therefore, the PMF condition will not necessarily be examined for structural stability.

f. Maximum design flood. The maximum design flood (MDF) is the designation used to represent the maximum structural loading condition (as judged by the minimum factor of safety) and must be determined for each structure or even for each structural element. MDF may be any event up to PMF. For floodwalls, MDF is usually when the water level is at or slightly above the top of the wall. Overtopping from higher water levels would result in rising water levels on the protected side, thus reducing net lateral forces. The same situation may be true for dams, but often significant overtopping can occur without significant increases in tailwater levels. The design engineer must consult with the hydraulics engineer to explore the possible combinations of headwater and tailwater and their effects on the structure. Some elements of dam outlet works (such as chute walls or stilling basins) are loaded differently from the main dam monoliths. For such elements, different flow conditions will produce maximum structural loading. When it is not obvious which loading will produce the lowest factor of safety, multiple loadings should each be investigated as a possible MDF. Since sliding is the most likely mode of failure for most gravity structures, MDF can usually be judged by determining maximum net shear forces. However, due to variable uplift conditions, a loading with smaller shears could result in the lowest factor of safety. Once the MDF is determined, it should be classified as usual, unusual, or extreme per Table 3-1, based on its return period.

4-4. Uplift Loads

Uplift loads have significant impact on stability. Sliding stability, resultant location, and flotation are all aspects of a stability analysis where safety can be improved by reducing uplift pressures. Since uplift pressures are directly related to flow paths beneath the structure, uplift pressure distribution may be determined from a seepage analysis. Such an analysis must consider the types of foundation and backfill materials, their possible range of horizontal and vertical permeabilities, and the effectiveness of cutoffs and drains. Techniques for seepage analysis are discussed in EM 1110-2-1901, EM 1110-2-2502, Casagrande (1937), Cedergren (1967), Harr (1962), and EPRI (1992). Seepage analysis techniques to determine uplift pressures on structures include flow nets, finite element methods, the line-of-creep method, and the method of fragments. Uplift pressures resulting from flow through fractures and jointed rock, however, are poorly understood and can only be accurately known by measurements taken at the point of interest. Joint asperities, changes in joint aperture, and the degree to which joints interconnect with tailwater influence uplift pressures and pressure distribution. Uplift pressures are site-specific and may vary at a given site due to changes in geology. Uplift pressures can be reduced through foundation drainage, or by various cutoff measures such as grout curtains, cutoff walls, and impervious blankets. Uplift pressures should be based on relatively long-term water elevations. Short duration fluctuations, such as from waves or from vibrations due to high velocity flows, may be safely assumed to have no effect on uplift pressures. Uplift pressures to be used for stability analysis of new structures are covered in Appendix C. The conservative uplift pressures used for the design of new structures may be significantly higher than those the actual structure may experience during its lifetime. For this reason, the use of actual uplift pressures for the evaluation of existing structures is permitted under the provisions discussed in Chapter 7. However, the engineer should be aware that in some instances the actual uplift may not be reflected by uplift cell readings. Since uplift measurement devices only capture a snapshot of a given part of the foundation, they should be used with caution, based on an overall evaluation of the foundation.

4-5. Maintenance Conditions

The return periods for a maintenance condition loading may be greater or less than 10 years, but based on past experience maintenance has been designated as an *unusual* load condition. The classification as an *unusual* loading is based on the premise that maintenance loadings take place under controlled conditions and that the structure performance can be closely monitored during maintenance.

4-6. Surge and Wave Loads

a. General. Surge and wave loads are critical in analyzing the stability of coastal protection structures but usually have little effect on the stability of inland structures. Wave and water level predictions for the analysis of structures should be based on the criteria presented in the Shore Protection Manual 1984, EM 1110-2-1612, and EM 1110-2-1614. Design forces acting on the structure should be determined for the water levels and waves predicted for the most severe fetch and the effects of shoaling, refraction, and diffraction. The methods recommended for calculation of wave forces are for vertical surfaces. Wave forces on other types of surfaces (sloping, stepped, curved, etc.) are not sufficiently understood to recommend general analytical design criteria. In any event, the structural engineer should consult with a coastal engineer in establishing wave forces for the design of critical structures.

b. Wave heights. Wave heights for design are obtained from the statistical distribution of all waves in a wave train and are defined as follows:

H_S = average of the highest one-third of all waves

$H_1 = 1.67\ H_S$ = average of highest 1 percent of all waves

H_b = height of wave which breaks in water depth d_b

c. Non-breaking waves. When the water depth is greater than approximately 1.5 times the wave height , waves do not break. The H_1 wave shall be used for the non-breaking condition. Design pressures for non breaking waves shall be computed using the Miche-Rudgren method. Whenever the maximum stillwater level results in a non-breaking condition, lower stillwater levels should be investigated for the possibility that shallow water may produce breaking wave forces, which are larger than the non-breaking forces.

d. Breaking waves. Waves break when the steepness of the wave and the bottom slope at the front of the structure have certain relationships to each other. It is commonly assumed that a wave will break if the water depth is not greater than 1.3 times the wave height. Study of the breaking process indicates that this assumption is not always valid. The height of the breaking wave and its breaking point are difficult to determine, but breaker height can equal the water depth at the structure, depending on bottom slope and wave period. Detailed determination of breaker heights and distances for a sloping approach grade in front of the structure are given in the Shore Protection Manual 1984. Design breaking wave pressure should be determined by the Minikin method presented in EM 1110-2-1614. Breaking-wave impact pressures occur at the instant the vertical face of the wave hits the structure and only when a plunging wave entraps a cushion of air against the structure. Because of this dependence on curve geometry, high impact pressures are infrequent against prototype structures; however, they must be recognized as possible and must be considered in design. Also, since the impact pressures caused by breaking waves are of very short duration, their importance in design against sliding and rotational instability may be questionable relative to longer lasting, smaller dynamic forces.

e. *Broken waves.* Broken waves are those that break before reaching the structure, but near enough to have retained some of the forward momentum of breaking. The design breaker height in this case (H_b) is the highest wave that will be broken in the breaker zone. Design wave forces for this height should be determined by the method presented in Chapter 7 of the Shore Protection Manual (1984).

4-7. Earthquake Loading Conditions

a. Seismic Load Conditions. Earthquake loads are used to represent the inertial effects attributable to the structure mass, the surrounding soil (dynamic earth pressures), and the surrounding water (hydrodynamic pressures). Design earthquakes shall comply with requirements of ER 1110-2-1806, based on the following seismic events.

- *Operational basis earthquake (OBE).* The OBE is considered to be an earthquake that has a 50 percent chance of being exceeded in 100 years (or a 144-year return period).

- *Maximum design earthquake (MDE).* The MDE is the maximum level of ground motion for which a structure is designed or evaluated. For *critical* structures the MDE is the same as the maximum credible earthquake (MCE). Generally, the probabilistically determined MDE for other structures is an earthquake that has a 10 percent chance of being exceeded in a 100-year period (or a 950-year return period).

- *Maximum Credible Earthquake.* The MCE is defined as the greatest earthquake that can reasonably be expected to be generated on a specific source, on the basis of seismological and geological evidence. The MCE is based on a deterministic site hazard analysis.

Earthquake-generated inertial forces associated with the OBE are unusual loads. Those associated with the MDE are extreme loads. Earthquake loads are to be combined with other loads that are expected during routine operations, and should not be combined with other infrequent events such as flood loads. Seismic loads should be combined with *coincident pool*, which is defined as the elevation that the water is expected to be at or below for half of the time during each year.

b. Analytical Methods. Several analytical methods are available to evaluate the dynamic response of structures during earthquakes: seismic coefficient, response spectrum, and time-history. These methods are discussed in reference ER 1110-2-1806. The current state-of-the-art method uses linear-elastic and nonlinear finite element time history analysis procedures, which account for the dynamic interaction between the structure, foundation, soil, and water. The seismic coefficient method, although it fails to account for the true dynamic characteristics of the structure-water-soil system, is accepted as a semiempirical method for determining if seismic forces control the design or evaluation, and to decide if dynamic analyses should be undertaken. The information in the following paragraphs describes the differences between the seismic coefficient method and dynamic analysis methods. Figure 4-2 illustrates the differences in the inertial and hydrodynamic earthquake loads obtained by the two different methods. The seismic coefficient used for the preliminary seismic stability evaluation of concrete hydraulic structures should be equal to 2/3 the effective peak ground acceleration (EPGA) expressed as a decimal fraction of the acceleration of gravity. The EPGA can be obtained by dividing the 0.30 second spectral acceleration, for the return period representing the design earthquake, by a factor of 2.5 The 0.30 second spectral acceleration is obtained from the spectral acceleration maps in Appendix D of ER 1110-2-1806.

c. Inertia force due to structure mass. In the seismic coefficient approach, the inertial force is computed as the product of the mass of the structural wedge (including the soil above the heel or toe and any water contained within the structure) and the seismic acceleration. This may also be expressed as the weight of the structural wedge times the seismic coefficient, expressed as a fraction of gravity.

$$F_h = m\,a = k_h\,W \qquad (4\text{-}1)$$

where: F_h = horizontal component of the inertial force (a similar equation can be used for vertical component)
m = mass of structural wedge
a = seismic acceleration
W = gross weight of structural wedge (including soil above the heel and toe, and water contained within the structure)
k_h = seismic coefficient = a / g
g = acceleration of gravity

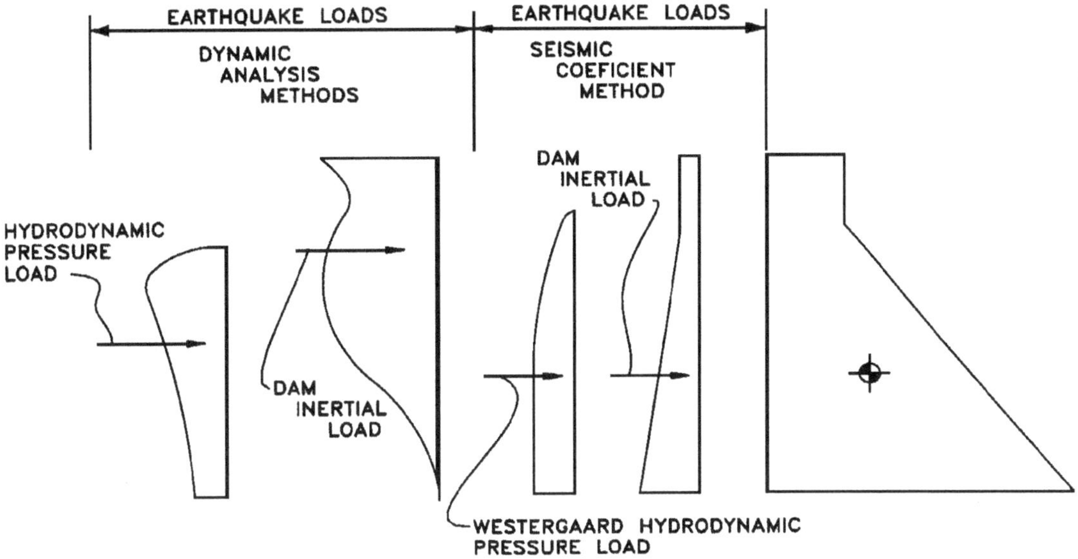

Figure 4-2 Idealized earthquake loads

The horizontal component of the inertial force is assumed to act at the center of mass of the structure, based on the assumption that the structure is a rigid body. In actuality, almost all structures have some flexibility, and the use of the rigid body concept often under estimates the magnitude of the inertial force. The location of the horizontal inertial force is also related to the flexibility of the structure, and usually acts at a location higher than the center of mass. However, because of the cyclic nature of earthquake loads, there is little probability of a rotational-stability related failure.

d. Inertial effects of soil. Backfill material adjacent to a structure will induce inertial forces on the structure during an earthquake. See Chapter 5 and Appendix G for information on soil loads due to earthquakes.

e. Effects of water. Water that is above the ground surface and adjacent to, or surrounding a structure will increase the inertial forces acting on the structure during an earthquake. The displaced structure moves through the surrounding water thereby causing hydrodynamic forces to act on the structure. The water inside and surrounding the structure alters the dynamic characteristics of the structural system, increasing the periods of the fundamental modes of vibration and modifying the mode shapes. In seismic coefficient methods, the hydrodynamic effects are approximated by using the Westergaard method (equation 4-2) (Westergaard 1933). The hydrodynamic force can either increase or decrease the water force, depending on direction of seismic acceleration. Figure 4-3 illustrates hydrodynamic pressures based on the Westergaard method.

$$P_E = (7/12)\, kh\, \gamma_w\, h^2 \qquad (4\text{-}2)$$

where: P_E = hydrodynamic force per unit length
k_h = horizontal seismic coefficient
γ_w = unit weight of water
h = water depth.

The hydrodynynamic force is added algebraically to the static water pressure force to get the total water force on the structure. The pressure distribution is parabolic and the line of action for the force P_E is 0.4 h above the ground surface. The Westergaard method assumes the structure is rigid and the water is incompressible. Since most structures are flexible, this method can lead to significant error. For free-standing intake towers, the hydrodynamic effects are approximated by adding mass to the structure to represent the influence of the water inside and surrounding the tower. Engineers using the seismic coefficient approach for stability analyses should be aware of the limitations and the simplifying assumptions made with respect to hydrodynamic pressures and their distribution on the structure.

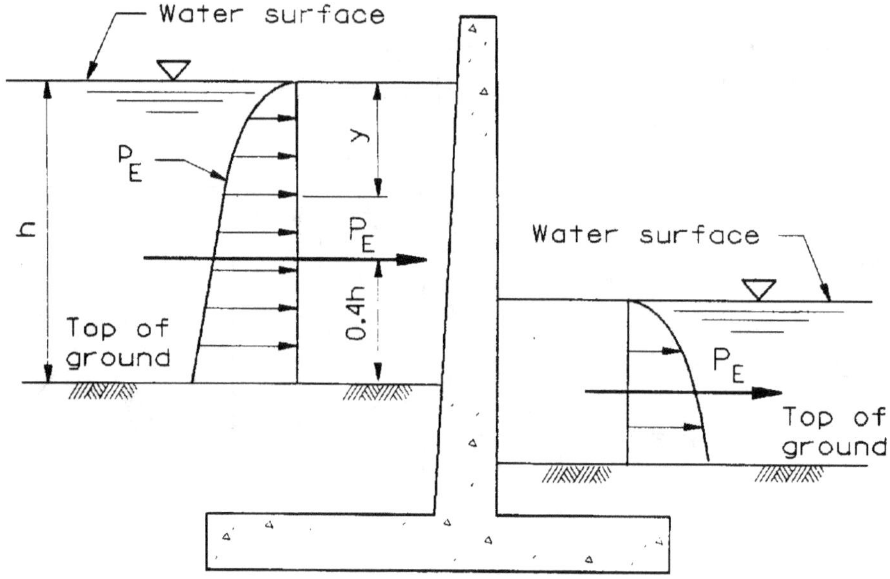

Figure 4-3 Hydrodynamic Forces for Freestanding Water

4-8. Other Loads

a. Impact. Impact loads for locks and dams on navigation systems are due to the structures being struck by barges. These loads can be quite large and for some structures, such as lock guide walls, control the stability analyses. Where impact loads must be considered, refer to EM 1110-2-2602.

b. Ice. Loads due to ice are usually not critical factors in the stability analysis for hydraulic structures. They are more important in the design of gates and other appurtenances. Ice damage to gates is quite common, but there is no known case of a dam failure due to ice. Where ice loads must be considered, refer to EM 1110-2-1612.

c. Debris. Debris loads, like ice loads, are usually of no consequence in stability analyses. However, they may be critical for the design of gates and floodwalls.

d. Hawser pull. Hawser pulls from barges are significant in the stability analysis for lock guide walls, mooring facilities, and floodwalls. Where hawser pulls must be considered, refer to EM 1110-2-2602.

e. Wind. Wind loads are usually small in comparison to other forces, which act on civil works structures. Therefore, wind loads should usually be ignored. For structures such as coastal flood walls where wind might cause instability, or for structures under construction, wind pressures should be based on the requirements of ASCE 7

f. Silt. Silt accumulation can occur upstream of dams. Not all dams will be susceptible to silt accumulation and the structural engineer should consult with hydraulic engineers to determine if silt buildup is possible, and to what extent it may accumulate over time. Silt loads should be included in the loading conditions indicated in Appendix B. Horizontal silt pressure is assumed to be equivalent to that of a fluid weighing 1362 kg/m^3 (85 pcf). Vertical silt pressure is determined as if silt were a soil having a wet density of 1922 kg/m^3 (120 pcf). These values include the effects of water within the silt.

4-9. Mandatory Requirements.

For a general discussion on mandatory requirements, see Paragraph 1-5. As stated in that paragraph, certain requirements within this manual are mandatory. The following are mandatory for Chapter 4.

a. Load Conditions. Stability shall be satisfied for all load conditions listed in Appendix B.

b. Maintenance. Maintenance load conditions shall be classified as *unusual.*

c. Water loading conditions. Water loadings shall be based on hydrologic analyses giving median water elevations in terms of return periods. Water elevations for various load conditions shall be as follows:

- *Coincident pool* shall be the elevation that the water is expected to be at or below for half of the time during each year. This water loading shall be used in combination with seismic loads.
- *Normal operation* loading shall represent maximum loads with a 10-year return period.
- *Infrequent flood* shall represent maximum loads with a 300-year return period.
- *Maximum design flood* shall be the maximum structural loading up to PMF.

d. Uplift loads. Uplift loads shall be calculated per the requirements of Appendix C, as follows:

- Uplift pressures shall be calculated based on an approximate seepage analysis and shall be applied over the full area of the base of the structure, or the failure plane under investigation.
- When a loss of contact is calculated to occur at the heel of the structure, full uplift pressure due to headwater shall be assumed to exist in this area. This provision does not apply to earthquake loading conditions.
- The maximum assumed effectiveness of drainage systems, cutoff wall systems, and combined drain and cutoff wall systems shall be 50%.
- Where overflow results in significant velocities and causes hydraulic jump and retrogression, tailwater pressures used in uplift calculations shall be reduced as described in Appendix C.

e. Earthquake loads.

- Earthquake loads shall be based on design earthquakes specified in ER 1110-2-1806.
- Structural inertia loads shall be calculated using the seismic coefficient method.
- Hydrodynamic loads shall be calculated using Westergaard's formula.
- Soil loads for seismic events shall be calculated per the requirements of Chapter 5.

Chapter 5
Soil Forces and Single-Wedge Sliding Analysis

5-1. General

Chapter 4 described various loading conditions and specific loads, except for soil loads. Chapter 5 describes soil loads and explains how to use various loads in a single-wedge stability analysis. The methods presented in this chapter are intended to produce reasonably conservative estimates of soil forces acting on a structure. This manual only addresses normal soil conditions, other conditions such as swelling soils require special studies. The definitions of terms that will be used throughout this chapter are as follows:

- *Single wedge.* The single wedge is the wedge to which forces are applied, i.e., the structure itself, which is referred to as the structural wedge.

- *Applied driving forces.* Driving forces are defined as those lateral forces whose primary influence is to decrease structural stability. The side of the structure upon which these forces are applied will be called the driving side. Uplift and downdrag are also treated as applied forces.

- *Applied resisting forces.* Resisting forces are defined as those lateral forces whose primary influence is to increase structural stability. The side of the structure upon which these forces are applied will be called the resisting side. The resisting side is on the opposite side of the structural wedge from the driving side. The difference between the driving and resisting forces is transferred to the foundation by the structural wedge.

- *Reactions.* The shear and the normal force between the foundation and the base of the single wedge are reactions, which are necessary to place the structure in static equilibrium, they are not included in the applied forces.

5-2. Single-Wedge Stability Analyses

a. Basic requirements. For the single-wedge analysis, the engineer must calculate the driving and resisting soil forces, lateral water forces, and uplift and apply them to the structural wedge. The vertical drag force discussed in Appendix F may also be included, if the requirements stated in the appendix are satisfied. Using these forces and the weight of the structural wedge, the engineer can determine the magnitude, location, and slope of the resultant acting at the base of the structural wedge. The resultant will be used to evaluate sliding stability, and the location of the resultant will be used to determine the potential for partial loss of contact between the structure and the foundation materials. The resultant and its location will also be used to evaluate the bearing capacity of the foundation materials. The forces applied to the structural wedge will also be used for design of the structural elements (e.g., shears, moments and axial loads in the base and stem of a retaining wall).

b. Soil forces. Lateral soil forces acting on the single wedge should be calculated using the minimum required factor of safety against sliding to obtain the developed soil strength parameters ϕ_d and c_d. The use of developed parameters results in a force larger than the active soil force on the driving side, and a force smaller than the passive soil force on the resisting side These developed parameters shall be used in Equations 5-3 through 5-15 to calculate the lateral soil forces acting on the driving side, and used in Equations 5-16 through 5-22 to calculate the lateral soil forces acting on the resisting side. Soil forces due to seismic events are discussed in Section 5-5. If there is a significant difference between the calculated and minimum required safety factors, it may be appropriate to re-evaluate the developed soil strength parameters used to determine the soil forces. The values for the developed soil strength parameters shall be determined as:

$$\phi_d = \tan^{-1}\left(\frac{\tan\phi}{FS_s}\right) \quad and \quad c_d = \frac{c}{FS_s} \qquad (5\text{-}1,\ 5\text{-}2)$$

where FS_s = required factor of safety against sliding

ϕ = nominal angle of internal friction of the soil

c = nominal cohesive strength of the soil

c. *Sliding.* The resultant can be resolved into components parallel and normal to the base plane of the structural wedge. The sliding factor of safety is calculated as follows:

$$FS_s = \frac{N \tan \phi + c\,L}{T} \tag{5-3}$$

where: N = the component of the resultant normal to the base

T = the component of the resultant parallel to the base

L = length of base in compression

If the safety factor is equal to or greater than the required safety factor, the sliding stability criterion is satisfied. Note that this calculated safety factor might not be equal to the minimum factor used to determine developed soil strength parameters. If there is a significant difference between the calculated and minimum required safety factors, it may be appropriate to re-evaluate the developed soil strength parameters used to determine the soil forces. The resultant location shall be used to determine the length of the base that is in compression, and cohesion shall not be effective on that part of the base that is not in compression. If there is any loss of contact, uplift forces should be re-evaluated. However, for seismic events, cyclic loading periods are so short that the time is not sufficient for uplift pressures to change, therefore, uplift should not be increased due to loss of contact for seismic loads.

5-3. Soil Pressures and Forces

a. Active soil pressures.

(1) Cohesionless backfill. Cohesionless materials such as clean sand are the recommended backfill for most structures. Large-scale tests (Terzaghi 1934; Tschebatarioff 1949; Matsuo, Kenmochi, and Yagi 1978) with cohesionless backfills have shown that lateral pressures are highly dependent on the magnitude and direction of wall movement. The minimum lateral pressure condition, or active soil pressure, develops when a structure rotates about its base and away from the backfill an amount on the order of 0.001 to 0.005 radians. As the structure moves, horizontal stresses in the soil are reduced, and vertical stresses due to backfill weight are resisted by increasing shear stresses until shear failure is imminent (Figures 5-1 and 5-2).

(2) Cohesive backfill. For situations where cohesive backfill is unavoidable, solutions are included herein for soil pressures involving both frictional and cohesive soil strength parameters (ϕ and c). Where cohesive backfill is used, two analyses (short-term and long-term) are usually required in order to model conditions that may arise during the life of the structure. Short-term analyses model conditions prevailing before pore water pressure dissipation occurs, such as the end-of-construction condition. Unconsolidated-undrained test parameters, which yield a relatively high cohesion value and a low or zero friction value, are appropriate for short-term analyses. Long-term analyses model conditions prevailing after shear-induced pore pressures have dissipated. For long-term analyses, consolidated-drained test parameters are appropriate. These tests usually yield a relatively high value for internal friction and a low or zero value for cohesion.

b. Passive soil pressures. If a structure is moved toward the backfill, lateral soil pressures increase and shear stresses reverse direction, first decreasing and then increasing to a maximum at failure (Figure 5-4). Full development of passive pressure requires much larger structure rotations than those required for the active case, as much as 0.02 to 0.2 radians (Figures 5-1 and 5-2). However, the rotation required to develop one-half of the passive pressure is significantly less, as little as 0.005 radians. The designer must be certain that soil on the resisting side of any structure will always remain in place and not be excavated or eroded before its effect is included in the stability analyses.

c. At-rest soil pressure. If no structural movement occurs, then the at-rest condition exists.

d. Design soil pressures - driving side. In practice, the active and passive soil pressure conditions seldom exist. Hydraulic structures are designed using conservative criteria that results in relatively stiff structures. Structures founded on rock or stiff soils usually do not yield sufficiently to develop active pressures. Even for foundations capable of yielding, experiments with granular backfill (Matsuo, Kenmochi, and Yagi 1978) indicate that following initial yield and development of active pressures, lateral pressures may in time return to greater values. Another reference (Casagrande 1973) states that the gradual buildup of the backfill in compacted lifts produces greater-than-active pressures, as do long-term effects from vibrations, water level fluctuations, and temperature changes. For these reasons and because large rotations are required for the development of passive pressures, soil pressures on both the driving side and the resisting side of the single wedge will be estimated by using the developed soil strength parameters, as defined in paragraph 5-2. These parameters are then used to calculate the *equivalent-fluid* soil pressure coefficients (K).

(1) General wedge method for equivalent fluid pressure coefficients. Lateral soil forces are assumed to act parallel to the top surface of driving side wedges when the surface slopes downward toward the structure. Equivalent fluid-pressure coefficients are calculated as follows:

$$\alpha = \tan^{-1}\left(\frac{C_1 + \sqrt{C_1^2 + 4C_2}}{2}\right) \tag{5-4}$$

where α = the critical slip plane angle for the soil wedge (see Appendix E for a derivation of α)

$$C_2 = \frac{t + \left(\dfrac{2V}{\gamma(h^2 - d_c^2)}\right)(1 + \tan^2\phi_d)\tan^2\beta + \left(\dfrac{2c_d}{\gamma(h + d_c)}\right)r}{A}$$

$$A = \tan\phi_d + \tan\delta - \left(\frac{2V}{\gamma(h^2 - d_c^2)}\right)(1 + \tan^2\phi_d) + \left(\frac{2c_d}{\gamma(h + d_c)}\right)r$$

$r = 1 - \tan\delta \tan\phi_d - \tan\beta (\tan\delta + \tan\phi_d)$

$s = \tan\beta + \tan\phi_d + \tan\delta (1 - \tan\beta \tan\phi_d)$

$t = \tan\phi_d - \tan\beta - (\tan\delta + \tan\beta) \tan^2\phi_d$

ϕ = soil internal friction parameter

ϕ_d = developed internal friction parameter

c = soil cohesion parameter

c_d = developed cohesion parameter

β = top surface slope angle, positive when slope is upward when moving away from the structure (When the top surface of the backfill is broken, solutions for α may be obtained by using analogous positive and negative strip surcharges.)

δ = wall friction angle. When β is positive, δ = β. When β is zero or negative δ = 0. Vertical shear (drag), as discussed in Appendix F, shall not be used to calculate the value of α or equivalent fluid soil pressure coefficients. However, drag may be used in addition to lateral soil pressures when the requirements of Appendix F are satisfied.

γ = average unit weight of soil (moist weight above water table, buoyant weight below)

V = strip surcharge

h = height of vertical face of soil wedge

d_c = depth of cohesion crack in soil (should always be assumed filled with water when calculating lateral forces)

The equivalent fluid-pressure coefficients are:

$$K = \frac{1 - \tan\phi_d \cot\alpha}{\cos\delta\,[(1 - \tan\delta\tan\phi_d) + (\tan\phi_d + \tan\delta)\tan\alpha]}$$

and for soils that possess cohesive properties as well as internal friction:

$$K_c = \frac{1}{2\cos^2\alpha(\tan\alpha - \tan\beta)[1 - \tan\delta\tan\phi_d + (\tan\delta + \tan\phi_d)\,tanalpha\,]}$$

The equation for the depth of a cohesive crack is:

$$d_c = \frac{2K_c c}{K\gamma\left(\dfrac{\tan\alpha}{\tan\alpha - \tan\beta}\right)}$$

And the total lateral soil force is calculated as:

$$P = \frac{1}{2}K\gamma\left(\frac{\tan\alpha}{\tan\alpha - \tan\beta}\right)(h - d_c)^2 + KV\tan\alpha$$

Examples are presented in Appendix D.

When any of the variables in the above equations are not present in a particular problem, they are set equal to zero, thereby simplifying the equations. Figure 5-3 illustrates a wedge containing cohesionless soil, showing the methods used to calculate the lateral force, and the soil pressure at any point on the vertical face of the wedge. Figure 5-4 shows similar information for a wedge consisting of a cohesive soil. Figure 5-5 shows the method used to determine the pressure distribution for a strip surcharge (a line load V).

When ß is greater than ϕ_d a solution for α cannot be obtained from Equation 5-4 because the number under the radical will be negative, making the square root indeterminate. However, when ß is equal to ϕ_d, Equation 5-4 will give a value for α equal to ß and ϕ_d. When $\alpha = \phi_d =$ ß, the total lateral soil force, for a granular soil not supporting a strip surcharge, is:

$$P = \frac{1}{2} \gamma h^2 \cos \phi_d$$

This equation gives the maximum driving side lateral soil force that can occur and should be used when ß is equal to or greater than ϕ_d.

(2) Equivalent fluid pressure coefficients from Coulomb's equation. Coulomb's equation may be used to calculate the equivalent fluid-pressure coefficient when the surface of the backfill wedge is planar and unbroken, if certain conditions are met. These conditions are:

- When backfill has a sloping top surface. There can be only one soil material. The water-table must be either completely above or completely below the backfill. The backfill must be cohesionless. Surcharges must be uniformly distributed and cover the entire top surface of the backfill wedge.

- When backfill has a horizontal top surface. There may be more than one soil material, if the top surface of all soil layers are horizontal. The water-table may lie within the backfill. The soil may be either cohesive or cohesionless. Surcharges must be uniformly distributed and cover the entire top surface of the backfill wedge.

When any of the above conditions are not applicable, the equivalent fluid pressure coefficient determined by the wedge method shall be used. The equation for the equivalent fluid-pressure coefficient, using the Coulomb Equation is:

$$K = \frac{\cos^2 \phi_d}{\cos \delta \left[1 + \sqrt{\frac{\sin (\phi_d + \delta) \sin (\phi_d - \beta)}{\cos \delta \cos \beta}} \right]^2}$$

and the total lateral soil force is:

$$P = \frac{1}{2} K \gamma h^2$$

(3) Equivalent fluid pressure coefficients for simple conditions. When the top surface of the wedge is horizontal, planar, supports a uniform surcharge covering the entire top surface, and the soil possesses cohesive strength as well as internal friction, the equivalent fluid pressure coefficients in the preceding equations reduce to the following simple expressions:

$$K = \frac{1 - \sin \phi_d}{1 + \sin \phi_d} = \tan^2 (45° - \frac{\phi_d}{2})$$

$$K_c = \sqrt{K}$$

$$d_c = \frac{2 c_d}{\gamma \sqrt{K}}$$

e. Design soil pressures - resisting side. Developed soil pressures on the resisting side may also be calculated using the developed soil strength parameters. In this manual, soil pressures and forces on the resisting side are generally assumed to act horizontally (wall friction angle $\delta = 0$). The equivalent fluid-pressure coefficient for soil pressures on the resisting side is calculated as follows:

$$\alpha = \tan^{-1}\left[\frac{-C_1 + \sqrt{C_1^2 + 4C_2}}{2}\right]$$

$$C_1 = \frac{2\tan^2\phi_d - \dfrac{4V}{\gamma h^2}[\tan\beta(1+\tan^2\phi_d)] + \dfrac{4c_d}{\gamma h}(\tan\phi_d - \tan\beta)}{A}$$

$$C_2 = \frac{\tan\phi_d(1+\tan\phi_d\tan\beta) + \tan\beta + \dfrac{2c_d(1+\tan\phi_d\tan\beta)}{\gamma h} - \dfrac{2V\tan^2\beta(1+\tan^2\phi_d)}{\gamma h^2}}{A}$$

where

$$A = \tan\phi_d + \frac{2c_d(1+\tan\phi_d\tan\beta)}{\gamma h} + \frac{2V(1+\tan^2\phi_d)}{\gamma h^2}$$

Then the equivalent fluid-pressure coefficients for resisting side pressures are:

$$K_P = \frac{1+\tan\phi_d\cot\alpha}{1+\tan\beta\tan\phi_d - (\tan\phi_d - \tan\beta)\tan\alpha}$$

and when cohesion is present:

$$K_{cP} = \frac{1}{2\cos^2\alpha(\tan\alpha - \tan\beta)[1+\tan\beta\tan\phi_d - (\tan\phi_d - \tan\beta)\,tanalpha]}$$

The developed soil force on the resisting side is then:

$$P_P = \frac{1}{2}K_P\gamma\left(\frac{\tan\alpha}{\tan\alpha - \tan\beta}\right)h^2 + 2K_{cP}c_d h + K_P V\tan\alpha$$

Since cohesive cracking will not occur on the resisting side, the term for the depth of cohesive cracking (d_c) is not applicable to these equations.

5-4. Soil Pressures with Water Table Within or Above Top of Backfill Wedge

Pressures and forces due to soil and water must be calculated separately, since wall friction (δ) is not applicable to water pressure and the equivalent fluid pressure coefficients (K and K_c) are for calculating lateral soil pressure only. K for water is always equal to one. However, the effective unit weight of soil below the water table is affected by the uplift due to water. In lateral soil-pressure calculations, the moist unit weight of soil is used above the water table, and the buoyant unit weight is used below the water table. When calculating lateral water pressure and uplift, the effect of seepage (if it occurs) must be considered. Lateral soil pressures and forces are calculated as shown below, and as illustrated in Figure 5-6. These must be added to the lateral water pressure.

$$P = \frac{1}{2} p_s (h - h_s) + \frac{1}{2} (p_s + p) h_s$$

Where: P = the total lateral soil force,

and $$p_s = K \gamma_m \left(\frac{\tan \alpha}{\tan \alpha - \tan \beta} \right) (h - h_s)$$ = lateral soil pressure at water table

$$p = K \left[\gamma_m h \left(\frac{\tan \alpha}{\tan \alpha - \tan \beta} \right) - (\gamma_m - \gamma_b) h_s \right]$$ = lateral soil pressure at bottom of soil wedge

γ_m = moist unit weight of soil, use above water table

γ_b = buoyant unit of weight of soil, use below water table

h = height of vertical face of soil wedge

h_s = height of water table above bottom of wedge

5-5. Earthquake Inertial Forces on Structures

The seismic response of structures with backfill is complicated because in addition to the interaction with the foundation, the structure is also subjected to the dynamic soil pressures induced by ground shaking. Dynamic backfill pressures are related to the relative movement between the soil and the structure and the stiffness of the backfill. Behavior of the structure may be controlled by rocking and/or translation in response to earthquake shaking. The type of response will correlate to different distributions of backfill pressure acting against the wall. The appropriate method for analyzing the backfill pressure may be categorized according to the expected movement of the backfill and structure during seismic events.

a. Behavior of Backfills.

(1) Non-Yielding Backfills. For low intensity ground motions the backfill material may respond within the range of linear elastic deformations. Walls with non-yielding backfills can be expected to have dynamic soil pressures greater than those predicted by the Mononobe-Okabe method. The dynamic soil pressures and associated forces in the backfill may be analyzed as an elastic response using Wood's method as described in ITL 92-11 (Ebeling and Morrision, 1992). A reasonable estimate for determining the additional lateral seismic soil force against a soil retaining structure, for non-yielding backfill conditions, can be determined as

$$F_{sr} = \gamma h^2 k_h$$

where: F_{sr} = Lateral seismic force representing dynamic soil pressure effects

γ = Unit weight of soil. Use moist or submerged unit weight when all soil is above or below the water table. For partially submerged soils the unit weight shall be proportioned by a weighted average.

h = Height of backfill

k_h = Effective peak ground acceleration, expressed as a decimal fraction of the acceleration of gravity

The seismic component of the total soil force F_{sr} is assumed to act at a distance of 0.63 h above the base of the wall. This force must be combined with the structure lateral inertial force, and if water is present, hydrodynamic seismic forces to obtain the total seismic force on the wall. Evaluation of a wall with non-yielding backfill for the aforementioned seismic forces is illustrated by Example 32 of ITL-92-11 (Ebeling and Morrison, 1992). The various seismic forces described above must be combined with static soil pressure forces and static water pressure forces to get the total force on the wall. Soil retaining structures not meeting stability criteria using the preliminary screening method should be evaluated using refined analysis techniques described in ITL-92-11 (Ebeling and Morrison, 1992).

(2) Yielding Backfills. The relative motion of the structure and backfill material may be large enough to induce a limit or failure state in the soil. This condition may be modeled by the Mononobe-Okabe method (Mononobe and Matsou (1929), and Okabe (1926)), in which a wedge of soil bounded by the structure and an assumed failure plane are considered to move as a rigid body with the same horizontal acceleration. The dynamic soil pressures using this method are described in Appendix G.

(3) Partially Yielding Backfills. The intermediate condition in which the backfill soil undergoes limited nonlinear deformations corresponds to the shear strength of the soil being partially mobilized. The dynamic backfill pressures may be estimated using an idealized constant parameter, SDOF model of a semi-infinite uniform soil layer (Veletsos and Younan 1994) or a frequency-independent, lumped parameter, MDOF system. The dynamic pressures for an irregular backfill may be analyzed using a soil-structure-interaction model such as FLUSH (Lysmer et al. 1975). The wall is usually modeled with 2-D elements. The foundation rock is represented by 2-D plane-strain elements with an appropriate modulus, Poisson's ratio, and unit weight. Transmitting boundaries in the form of dashpots are introduced at the sides of the foundation rock to account for the material nonlinear behavior with depth. The shear modulus and soil damping vary with the level of shearing strain, and this nonlinear behavior is usually approximated by an equivalent linear method. The boundary conditions for the backfill may also be represented by dashpots. Hydrodynamic pressures exerted on the wall are computed using the Westergaard formula.

b. Simplified Wedge Method. A seismic coefficient method may be used to estimate the backfill and wall inertial forces acting on a single wedge. Theoretically, a structure (wedge) may behave as a rigid body that is fully constrained along its base and sides by the ground, so all parts of the wedge would be uniformly affected by accelerations which are identical to the time history of the ground motions. Therefore it would be appropriate to use a seismic coefficient equal to the peak ground acceleration for stability analysis of short, stiff structures. However, field and test data show that most structures do not behave as a rigid body, but respond as a deformable body subjected to effective ground motions. Thus the magnitude of the accelerations in a deformable wall may be different than those at the ground surface, depending on the natural period and damping characteristics of the structure and the shaking characteristics of the ground motions. Furthermore the maximum acceleration will only affect the structure for a short interval of time, and the inertia forces will not be equivalent to those of an equal static force which would act for an unlimited time, so the deformations resulting from the maximum acceleration will be smaller. Design or evaluation of structures for zero relative displacement under peak ground accelerations is unrealistic, so the seismic stability analysis should be based on a seismic coefficient, which recognizes that an acceptably small amount of lateral displacement will likely occur during a major earthquake. Experience has shown that a seismic coefficient equal to 2/3 of the peak ground acceleration is a reasonable estimate for many hydraulic structures. For partially yielding backfill, the strength mobilization factor should be equal to the reciprocal of the minimum required sliding safety factor for that load case. More information about the simplified wedge method is included in appendix G.

5-6. Mandatory Requirements

For a general discussion on mandatory requirements see Paragraph 1-5. As stated in that paragraph, certain requirements within this manual are mandatory. The following are mandatory for Chapter 5.

a. Developed Soil Strength Parameters. Lateral soil forces acting on a single wedge shall be determined using developed soil strength parameters as described in paragraph 5-2.b.

b. Sliding Factor of Safety. The factor of safety for sliding shall be calculated as defined in paragraph 5-2.c.

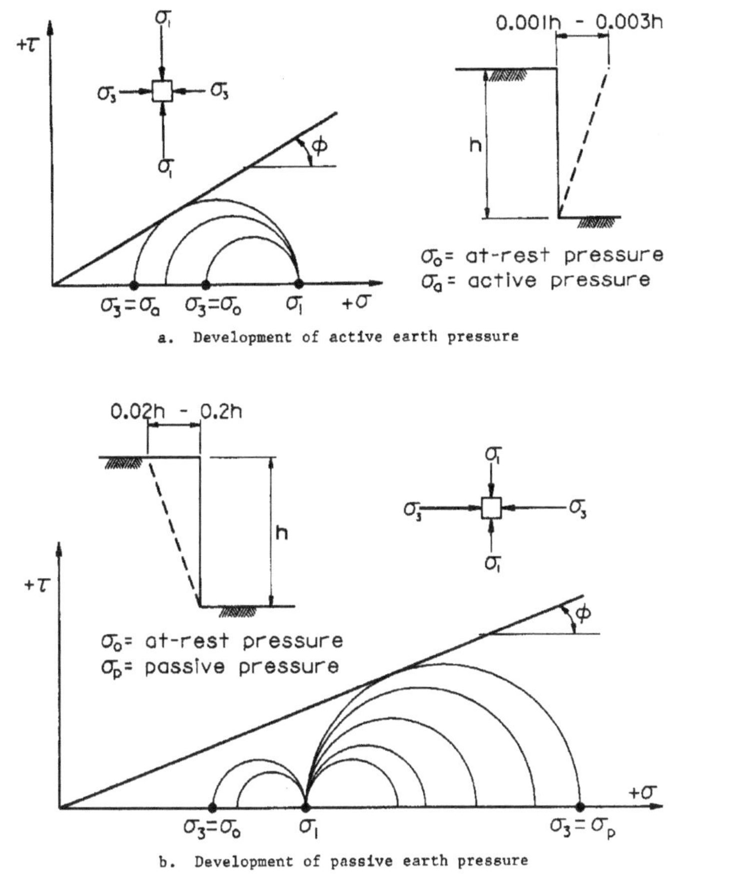

Figure 5-1 Soil Pressures for Cohesionless Material

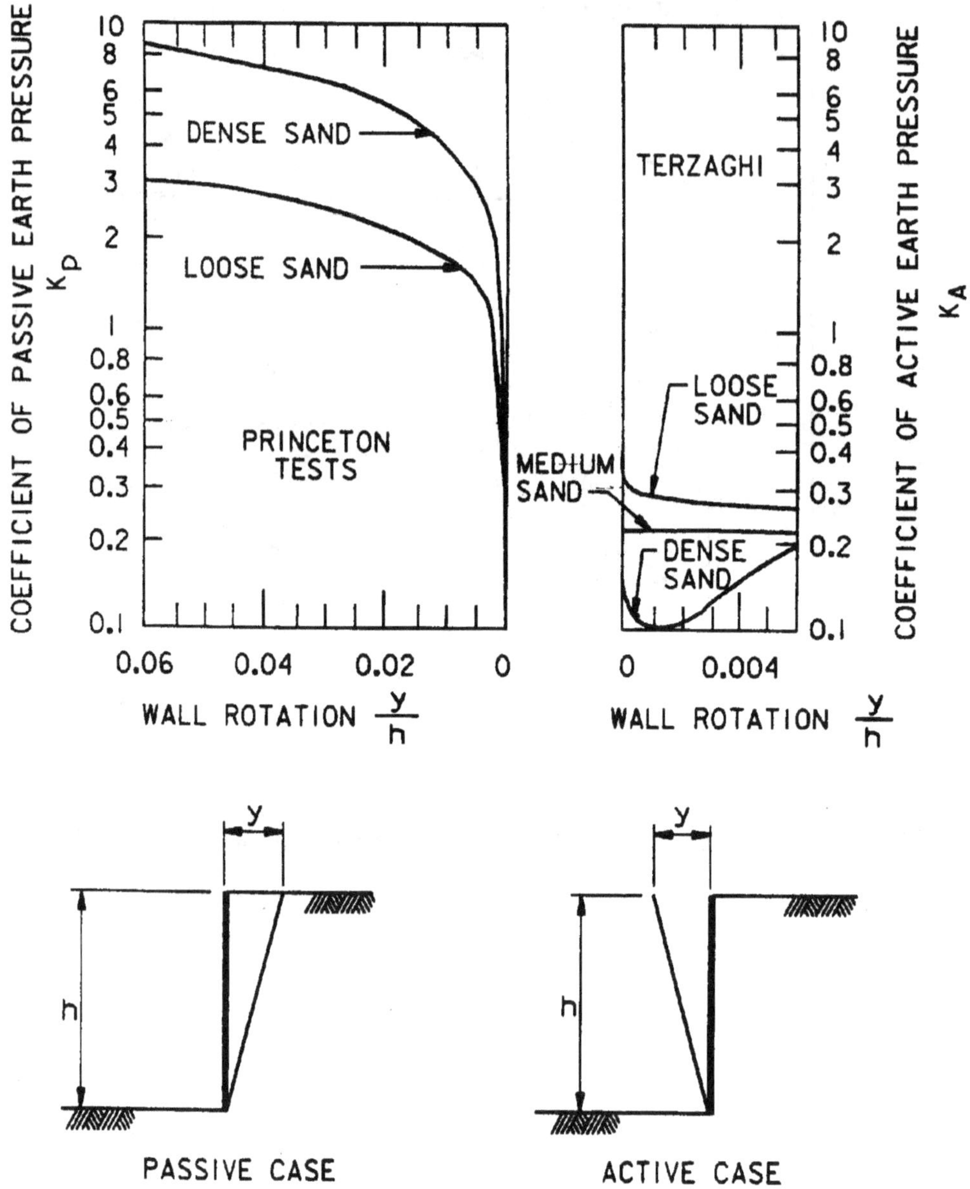

Figure 5-2 Soil Pressures versus Wall Movements

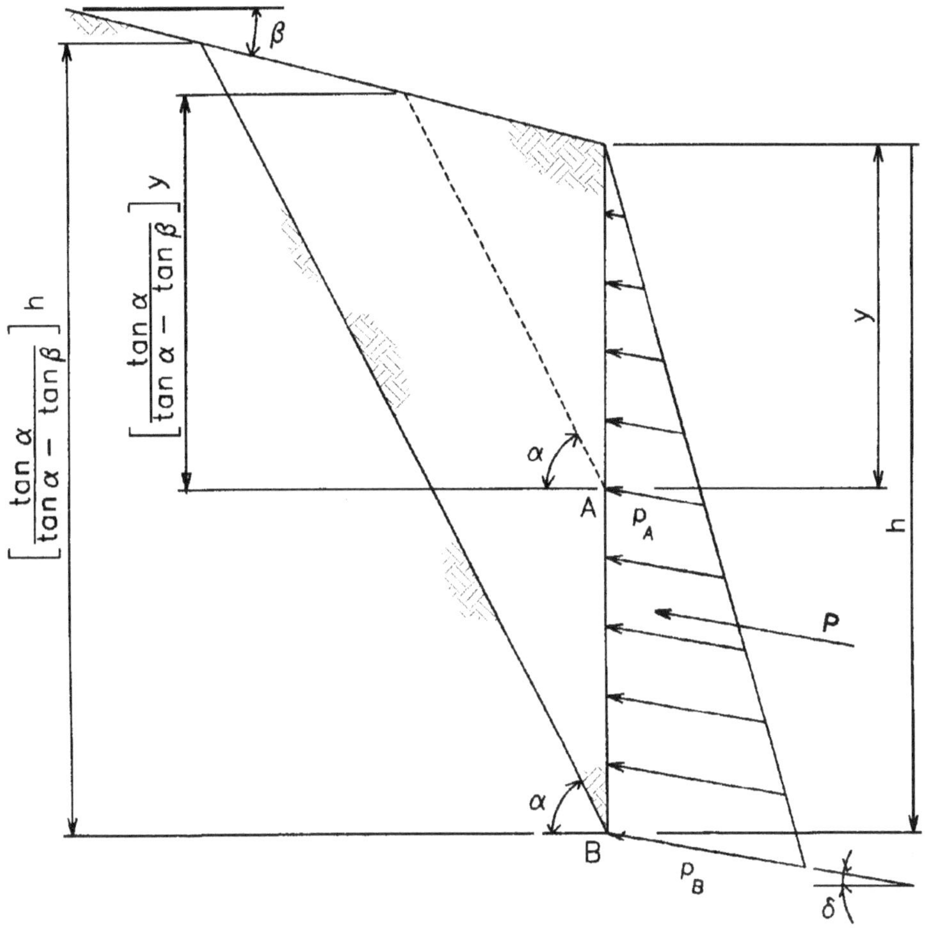

$$p_A = K\gamma\left[\frac{\tan\alpha}{\tan\alpha - \tan\beta}\right]y \qquad p_B = K\gamma\left[\frac{\tan\alpha}{\tan\alpha - \tan\beta}\right]h$$

$$P = 1/2\, K\gamma\left[\frac{\tan\alpha}{\tan\alpha - \tan\beta}\right]h^2$$

Figure 5-3 Calculation of Soil Pressures for Cohesionless Material

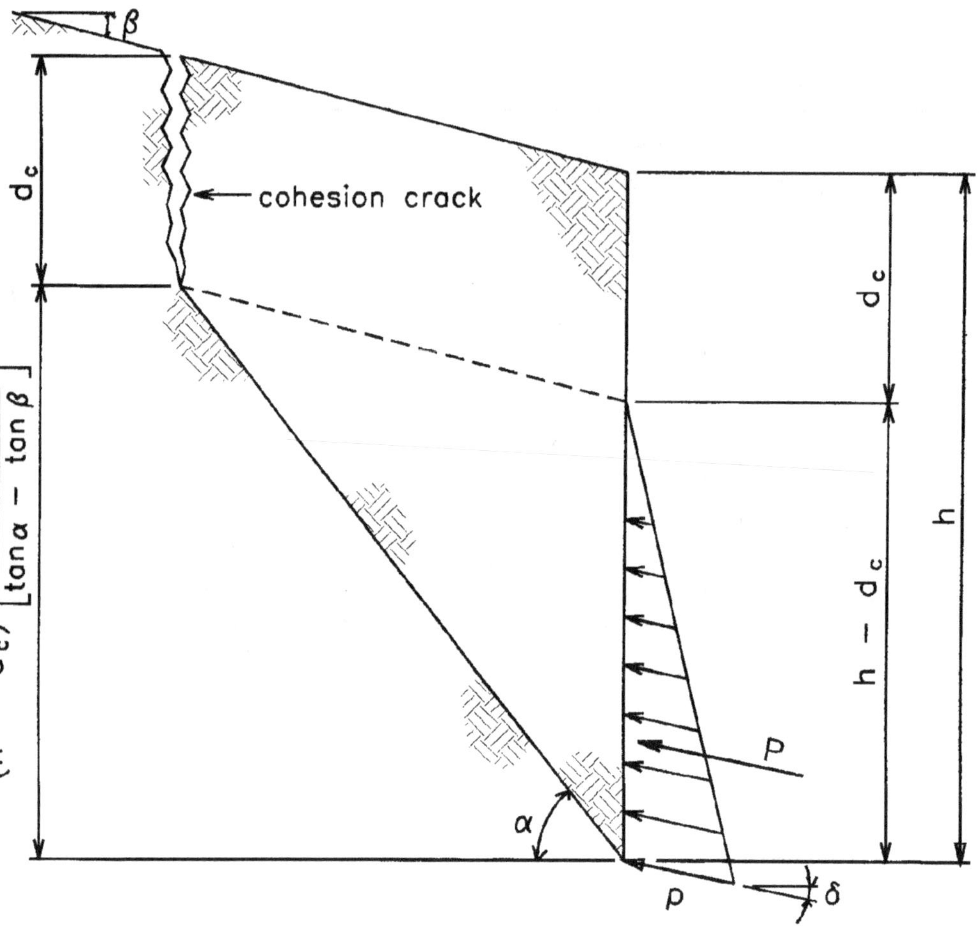

$$p = K\gamma\left[\frac{\tan\alpha}{\tan\alpha - \tan\beta}\right](h - d_c)$$

$$P = 1/2\ K\gamma\left[\frac{\tan\alpha}{\tan\alpha - \tan\beta}\right](h - d_c)^2$$

Figure 5-4 Calculation of Soil Pressures for Cohesive Material

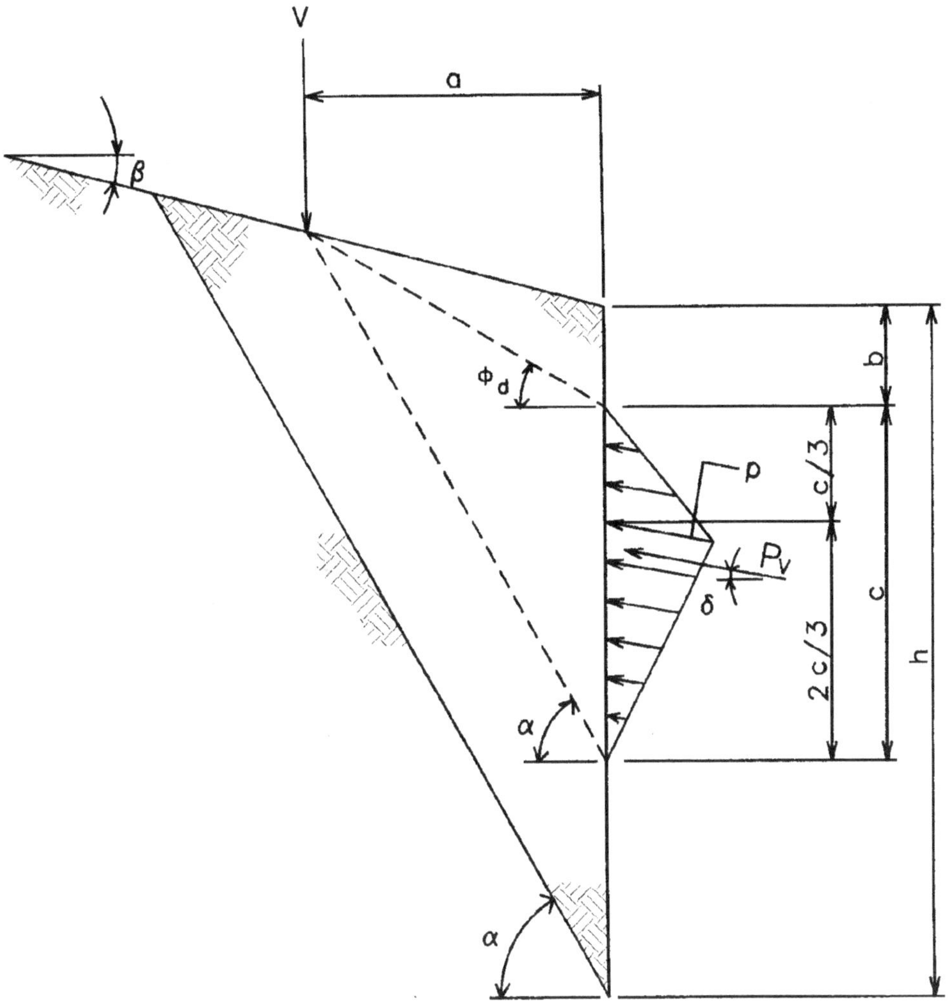

$$b = a\,(\tan\phi_d - \tan\beta)$$

$$c = a\,(\tan\alpha - \tan\phi_d)$$

$$P_V = K\,V\,\tan\alpha$$

$$p = \frac{2P_V}{c}$$

Figure 5-5 Lateral soil pressures due to strip surcharge loads

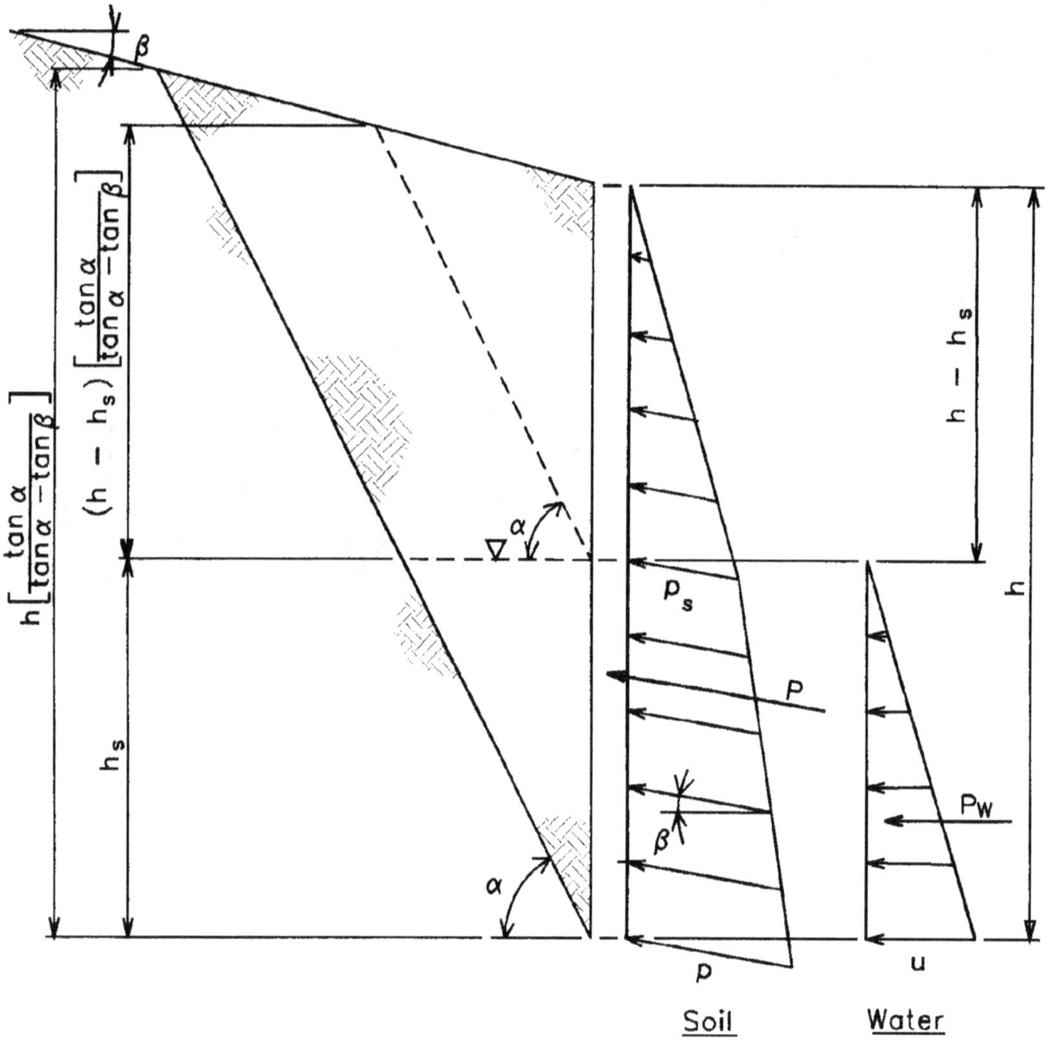

Figure 5-6 Lateral Soil Pressures with Water Table

Chapter 6
Stability Considerations and Analytical Methods

6-1. General

Both new and existing structures are constantly being examined to determine if they meet stability criteria. The traditional procedure is to evaluate sliding stability of these structures using the limit equilibrium method of analysis and to evaluate rotational stability using rigid body assumptions and bearing pressure distributions that vary linearly across the base. Soil and uplift loads on the structure are often based on simplifying assumptions. In the limit equilibrium method, only the stress state at failure is considered, usually as represented by the Mohr-Coulomb limit-state criterion. The simplified loads and the Mohr-Coulomb limit-state criterion are then used to obtain a single overall factor of safety against a sliding stability failure. This traditional approach is appropriate for many structures where it is difficult to predict just how a particular failure mechanism may develop. However, for existing structures where more is known about service state conditions and uplift pressures, it is possible to more precisely predict what stress changes must occur to develop a failure condition. The margin of safety can be more accurately determined when the path to failure, from initial stress conditions to limit state conditions, is known. This path to failure can be investigated using linear elastic and nonlinear finite-element numerical solutions and fracture mechanics concepts. Because the deformation of the structure and its foundation is considered in finite element analyses, and because the path to failure can be more realistically characterized through fracture mechanics, a more accurate assessment of safety can often be made using advanced analytical methods.

6-2. Traditional Methods

Traditional two-dimensional and three-dimensional, simplified, conservative stability analyses are used as the first step of a stability evaluation. Advanced analytical methods may be used to further investigate stability when the traditional methods indicate the structure is inadequate or marginal with respect to sliding stability or rotational stability.

a. *Two-dimensional analysis.* Two-dimensional stability analyses using traditional methods are suitable for the intermediate monoliths of structures that have uniform geometry and loading for their entire length and where the loads and resistance due to end effects are not transmitted across monolith joints. Many walls have variable height or loads along their length. Each monolith of such walls should be proportioned for stability at the low-end monolith joint, at the midpoint between monolith joints, and at the high-end monolith joint; the dimensions at different points along the wall should vary linearly between the sections for which the stability is checked.

b. *Three-dimensional analysis.* Some structures can not be suitably idealized as a two-dimensional structural system and therefore require a three-dimensional analysis. Examples of such structures include: nonoverflow monoliths at the ends of gravity dams, navigation-lock miter-gate monoliths, arch dams, and other structures having nonuniform geometry and/or loading. These types of structures, however, may still be analyzed using traditional methods with the following additional considerations.

(1) End effects. Figure 6-1 shows a plan and elevation of a short wall and the earth wedges that would be associated with sliding failure. However, frictional drag forces will exist on the active (driving) wedge faces abc and a'b'c', and on the passive (resisting) wedge faces edf and e'd'f' as well on the embedded end areas of the wall. These forces are generated by at-rest soil pressure on the end faces of the wedges and the wall, acting in conjunction with the internal soil friction. The frictional drag forces may be added to the numerator of Equation 5-1. When this resistance is used in conjunction with a multiple-wedge sliding analysis, as discussed in Chapter 2, $Ø_d$ (the developed value) should be used in lieu of Ø in calculations.

(2) Arch dams. Unlike a gravity dam, where stability is provided by the weight of the structural wedge, an arch dam's stability depends, not only on weight but to a much larger extent, upon arch action that transmits the imposed loads to the valley walls. Therefore, an evaluation of the foundation characteristics is required prior to laying out the dam. Following this procedure, the dam is designed so that the loads transmitted to the foundation will be within acceptable limits. In special instances artificial abutments (thrust blocks) may be used to increase the ability of the valley walls to resist these thrusts. Arch dams are constructed using monoliths in the same manner as

for gravity dams. The construction blocks or monoliths of double curvature arch dams may have a tendency to rotate upstream prior to completion. This tendency should be evaluated and a system of support should be evaluated, designed and included in the construction documents. The stability of these vertically cantilevered monoliths must be maintained at all stages of construction. EM 1110-2-2201 provides detailed information on site selection, stability requirements, design and construction criteria and procedures, methods of static and dynamic analysis, temperature studies, concrete testing requirements, foundation investigation, and instrumentation for use in the design of arch dams.

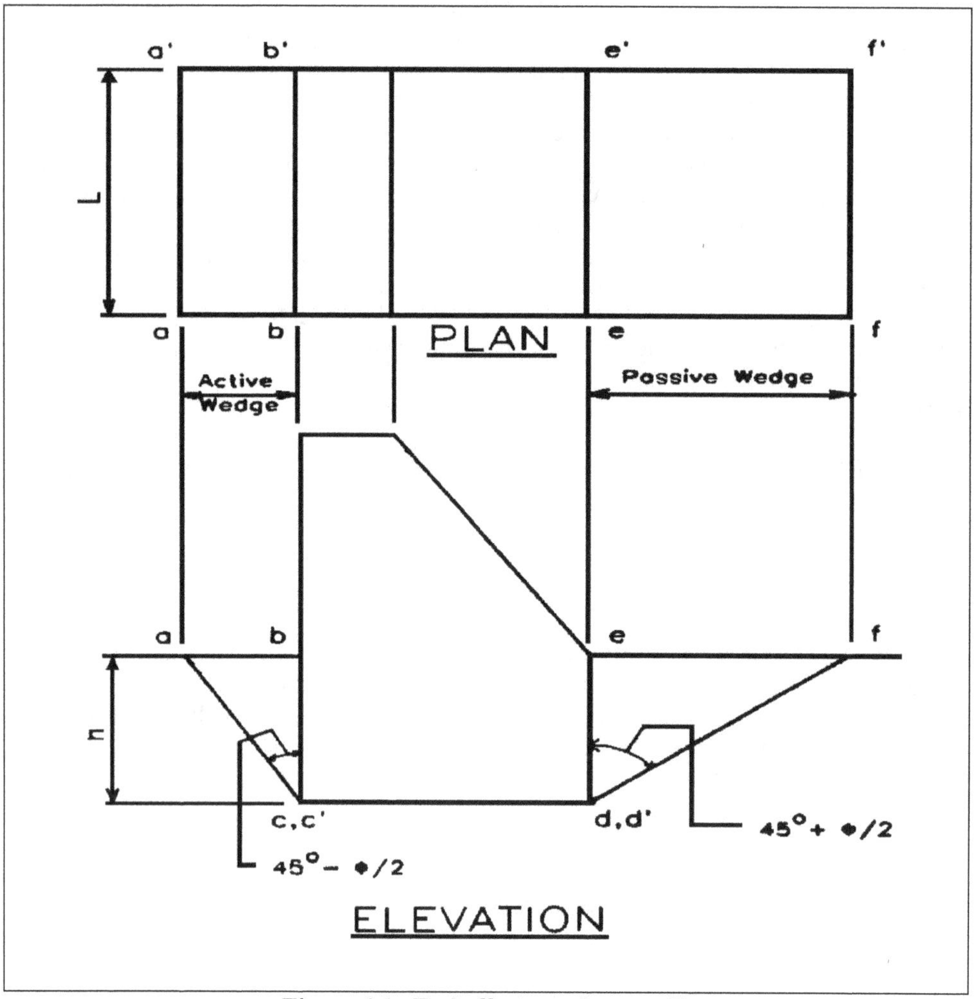

Figure 6-1. End effects on short wall

6-3. Advanced Analytical Analyses

Advanced analytical analyses can be used to gain a better understanding of structural behavior. However, the results obtained from such an analysis must be used in conjunction with the traditional analysis methods. The advanced analytical methods briefly described in this manual can be extremely difficult to perform and evaluate. Therefore, advanced analytical analyses should be conducted in conjunction with experts who can oversee the numerical modeling and can help with the interpretation of the results.

a. Stress/strain compatibility. Foundations may contain more than one material or may be made up of a combination of intact rock, jointed rock, or sheared rock. The stress-strain characteristics of these materials may be quite different. They might exhibit elastic-plastic behavior where peak shear strengths will occur simultaneously (Figure 6-2) or one or more might exhibit strain-softening behavior where only the residual strength of that material will be available at the strain level associated with peak shear in the other material (Figure 6-3). The overall strength of the foundation will depend on the stress-strain characteristics and compatibility of the various materials.

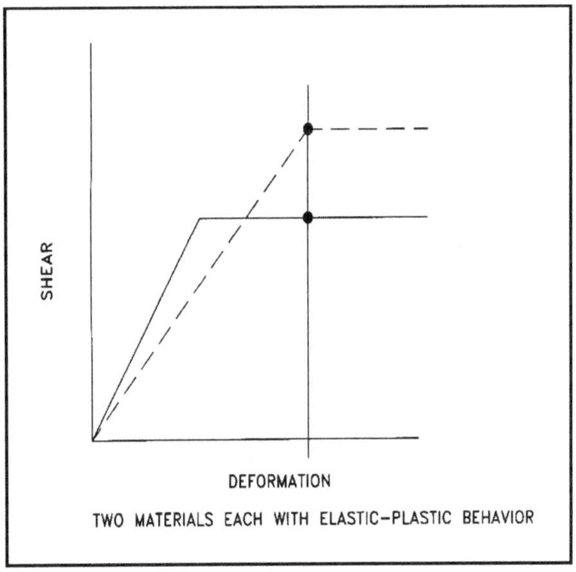

Figure 6-2

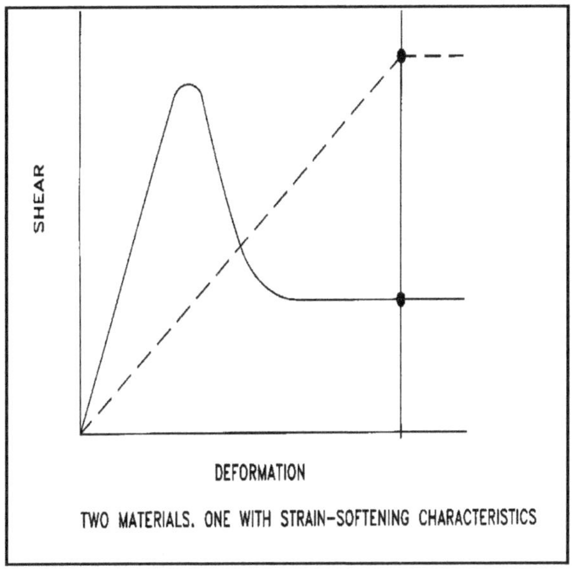

Figure 6-3

b. Deformations. The sliding stability analysis procedures described in Chapter 2 are based on a limit state approach, which satisfies only some of the equilibrium conditions and only accounts for deformations indirectly. No attempt has been made in this approach to relate shear stresses to displacements. The magnitude of displacement required to reach a limit state needs to be considered when selecting strength parameters. It is not possible, for instance, to develop peak shear and cohesive strength as well as passive resistance, all at the same strain level. Therefore, the analysis may require the use of residual shear strengths rather than peak shear strength values. In addition, the deformations associated with shear resistance may be unacceptable for reasons of serviceability. Controls on maximum displacement, needed to ensure proper function and safety, may govern over factor of safety requirements.

c. Special Circumstances. The minimum factors of safety for sliding stability are based on Mohr-Coulomb limit-state failure criterion without consideration of the deformations that occur in the structure and its foundation, or of the type of failure path the structure will experience in moving from a service-state from a limit-state condition. In many cases, it can be demonstrated by the use of finite element analysis and fracture mechanics methods that the factor of safety is substantially different from that predicted by the limit equilibrium method. Under special circumstances, such as in the case of an existing structure where the limit equilibrium method indicates that remedial action is required to improve stability, additional stability analyses may be performed using finite element and fracture mechanics procedures to verify whether stability remediation is actually required.

d. Finite element method (FEM). Finite element methods can be used to determine the manner in which loads and resistance are developed as a function of the stiffness of the foundation, stiffness of the structure, and the structure-to-foundation interface. They can also be used to calculate displacements and stresses due to incremental construction and/or load applications, and can model nonlinear stress-strain material behavior. The FEM program SOILSTRUCT, a two-dimensional plane strain-analysis program, is commonly used for soil-structure interaction problems and has been modified for use in evaluating the loss of contact along a crack, or along the structure-foundation interface. The modified version uses a procedure called the Alpha Method (Ebeling et. al 1992) in an incremental analysis to determine the extent of cracking that might occur at the structure-foundation interface. Finite element analyses can also be used to capture soil-structure interaction effects and to investigate crack propagation using fracture mechanics techniques. Soil-structure interaction. When a structure is supported by rock or soil, a finite element analysis can provide lateral pressures and bearing pressures that are realistic since the elastic and plastic properties of the structure, foundation, and soil are all considered.

e. Linear elastic fracture mechanics (LEFM) analysis. LEFM methods are FEM-based procedures for modeling crack development in a structure or along the base of a structure. The analysis can be performed using discrete crack analysis theory, or smeared crack analysis theory. The use of LEFM analysis in the evaluation of stability is described in various engineering papers (Dewey, Reich, and Saouma 1994), (Saouma, Bruhwiler, and Boggs 1990), (Bazant 1990) and (Ebeling, Morrison, and Mosher 1996).

6-4. Computer Programs

Following is a listing and brief description of computer programs that may serve as aids when determining the stability requirements for new or existing structures. Other commercially available computer programs which are not included in the following list, such as ADINA, ANSYS, CRISP, NASTRAN, or PLAXIS may also help when performing complex stability analysis.

- 3DSAD is a U.S. Army Corps of Engineers developed program (library number X8100) that allows the user to describe the geometry of a three-dimensional structure, interactively plot the described structure, and compute weight and centroid information for individual pieces or the sum total of the structure.

- CFRAG is a U.S. Army Corps of Engineers developed program (I0018) that allows the user to analyze groundwater flow using the method of fragments. The program can be used to compute: (a) seepage through soil mediums that can be modeled using fragment, (b) head losses, (c) exit gradients, and (d) resultant uplift and lateral forces.

- CSLIDE is a U.S. Army Corps of Engineers developed program (X0075) that allows the user to assess the sliding stability of concrete structures using the limit equilibrium method described in Chapter 2.

- CSEEP is a U.S. Army Corps of Engineers developed program (X8202) that allows the user to (a) interactively generate a finite element grid, (b) perform a finite element method seepage analysis, and (c) to plot the results.

- MERLIN is an Electric Power Research Institute (EPRI) developed program that allows the user to solve problems in elasticity, plasticity, linear and nonlinear fracture mechanics, steady-state and linear-transient heat transfer, and transient seepage flow. This program includes numerous capabilities for performing fracture mechanics analyses using the discrete crack approach. Pre- and post-processing capabilities are contained in PreMERLIN and PostMERLIN, respectively.

- SOILSTRUCT is a U.S. Army Corps of Engineers developed program that allows the user to perform a two-dimensional plane-strain finite element analysis of incremental construction, soil-to-structure interaction analysis of earth-retaining structures. The program is capable of simulating incremental construction including embankment construction or backfilling, the placement of layer(s) of a reinforcement material during backfilling, dewatering, excavation, installation of a strut or tie-back anchor support system, removal of the same system, and the placement of concrete or other construction materials. The program is also capable of modeling the interface region between the soil backfill and the structure using interface elements.

- CG-DAMS is an EPRI developed program that is a specialized nonlinear, finite element, concrete gravity-dam stability code that predicts cracking potential under all types of loads including extreme flood and earthquake loads. The code has library models of typical dam and foundation cross-sections that can be customized by entering elevations and slopes of the actual dam. Uniquely shaped dam profiles are modeled by user input. Output routines provide the sliding stability factor of safety, structural deformations, stress and pressure profiles in the dam foundation, and crack length.

Chapter 7
Evaluating and Improving Stability of Existing Structures

7-1. General

The stability of existing structures is sometimes reevaluated due to project modifications, changes in site conditions, improved knowledge of site data, or changes in stability criteria (ER 1110-2-1155). Since modifications to improve stability are often expensive, each structure should undergo a systematic, phased evaluation process to determine whether remediation is necessary. To avoid unnecessary modifications, all types of resisting effects, including those described below, should be considered. These include vertical friction, side friction, or three-dimensional effects. It may not be necessary to modify an existing structure that does not satisfy the requirements for new structures, especially when the structure's remaining life is short (scheduled to be replaced or removed) or when there are no indications of any stability problem. Where current stability requirements cannot be satisfied, waivers may be obtained subject to approval by CECW-E.

7-2. Procedures

The following procedures are to be used in the evaluation of existing structures and in the determination of necessary corrective measures to improve stability. The procedures should be considered as guides and are not intended to replace engineering judgment by the project engineers. The stability condition should always be reviewed when there are significant changes in the loading conditions, severe damage due to accident, aging or deterioration, discovery of structural deficiencies, revisions of stability criteria to become more conservative, or when required by ER 1110-2-100. The phases listed below shall be followed in sequence until the prescribed performance objectives described in Chapter 3 are satisfied. Existing structures can qualify for use of the *well-defined* site information category. This will permit use of lower factors of safety when evaluating sliding stability.

a. PHASE I; Preliminary analysis and evaluation. Preliminary analysis should be performed based upon available data and actual conditions of the structure. Before performing the analysis, collect and review all the available data and information for the structure including geologic and foundation data, design plans, as-built plans, periodic inspection reports, damage reports, plans of previous modifications to the structure, measurements of movement, instrumentation data, and other pertinent information. It may be necessary for the engineer to inspect and examine the existing structure to assess its condition. The first step in evaluating the stability of an existing structure is to perform a preliminary two-dimensional stability analysis. Uplift pressures, the angle of internal friction, and cohesion should be evaluated using parametric studies to assess the impact of each parameter on the factors of safety. If the results of the Phase I preliminary analysis indicate that the structure meets the safety and performance objectives of Chapter 3, no further investigations are required. Otherwise, various options to meet safety and performance objectives should be investigated.

b. PHASE II; Study, investigation, and comprehensive analyses. When the preliminary analysis indicates the structure does not meet safety and performance objectives, a plan of action for a comprehensive stability evaluation should be developed. The plan should determine the extent of the explorations and testing program needed to accurately define foundation strength parameters, the analytical program needed to accurately define the loading conditions, the remedial schemes to be studied, and the extent of any additional parametric studies. An exploration, sampling, testing, and instrumentation program should be developed, if needed, to determine the magnitude and the reasonable range of variation for the parameters which have significant effects on the safety and performance of the structure as determined by parametric studies. Testing laboratories should be used to the maximum extent practicable to develop shear strength data for the foundation and backfill materials. Comprehensive stability analyses should be performed using the material properties and strength information obtained from the sampling and testing program. Consideration may be given to reducing backfill pressures to active-state conditions provided the displacements required to reach an active-state condition meet the performance objectives in Chapter 3. Remedial measures will not be required to improve stability should the parametric studies conducted in Phase II indicate the safety and performance objectives of Chapter 3 can be met without remediation.

c. PHASE III; Advanced analytical studies and reliability analyses.

(1) Advanced analytical studies. If the stability of the structure is still in question after completing preliminary and comprehensive analyses, advanced analytical studies should be performed. These studies should use two- and three-dimensional finite element methods to capture the interaction between the foundation, backfill, and the structure, and to capture the capacity of the structural system to distribute loads to adjacent monoliths and abutments. A fracture mechanics analysis (Dewey, Reich, and Saouma, 1994) may be necessary to evaluate cracking and the progressive development of uplift pressures along potential failure planes. However, nonlinear analyses are not simple to perform, and fracture mechanics computer programs are not readily available. Fracture mechanics analysis should only be undertaken in special circumstances.

(2) Reliability analyses. Reliability analyses provide a alternate means of evaluating the stability of existing structures. This process combines what is known about a structure, with reasonable limits for the unknowns, to aid the engineer in assessing in probabilistic terms the potential for a stability failure. Although generally used for making investment decisions about projects competing for major rehabilitation funds, the process can be used to evaluate the risks associated with structures that, when analyzed by conventional deterministic procedures, indicate a stability-failure potential.

(3) Vertical shear. Many existing gravity earth-retaining structures have performed satisfactorily for many years although they do not meet stability criteria. Several research programs have addressed the stability of existing structures. The results are provided by Ebeling, Duncan, and Clough (1990), Filz and Duncan (1992), and Ebeling, Pace, and Morrison, (1997). This research demonstrated that, for rigid gravity walls on rock, where the rotation of the walls is not sufficient to fully mobilize the shear resistance of the backfill, a vertical shear force (drag) will develop on the back of the wall as a result of vertical settlement within the backfill. This shear force helps stabilize gravity walls founded on rock and should be considered when evaluating existing structures. The vertical shear force can be determined by a soil-structure interaction analysis using nonlinear finite element programs such as SOILSTRUCT as part of the Phase III analysis. Additional information and details on the use of vertical shear in stability analysis is presented in Appendix F.

7-3. Improving Stability

Stability can be improved through several means such as by the use of foundation drains to reduce uplift, by adding anchors, and by adding concrete mass, concrete buttresses, or buttressing embankments.

a. Reducing uplift pressures. In most cases, the pressure reduction provided by foundation drains is much greater than what is usually assumed for design. However, experience has shown that drains can clog, and this sometimes is temporarily unnoticed. If drains are properly maintained and if foundation testing and flow analysis provide supporting justification, the drain effectiveness can be increased beyond 50 percent subject to approval from CECW-E. This criterion will depend on the pool operational plan, instrumentation to verify and evaluate uplift assumptions, and an adequate drain maintenance program. When actual uplift pressures are known, these uplift pressures can be extrapolated to pool levels used for the stability evaluation. However, extrapolation of existing uplift readings at low-pool elevations for use under high-pool conditions must be done with caution and with full consideration of the effect joint aperture changes can have on high-pool uplift pressures and their distribution from upstream to downstream within the foundation. Joints in the foundation rock can open or close as pool conditions change and the joint aperture change may not be uniform. Also, joint aperture changes can result in something other than linear uplift pressure distribution from upstream to downstream within the foundation. For these reasons, the use of existing uplift pressures in stability analyses must be supported with uplift pressure readings that reflect uplift pressure changes that have occurred under various pool levels and various climatic conditions. Information on the influence that geology, foundation treatment, drainage, changes in reservoir elevation, and seasonal temperature variations have on uplift pressures can be found in EPRI (1992). Flood-pool conditions that cause tension at the heel of the structure must be checked to determine if a crack will develop between the structure and foundation, resulting in full uplift pressures in the crack. Uplift records must be sufficient to adequately define distributions across the base from headwater to tailwater, and cover as many years as necessary to discern if there is a potential for any uplift pressure changes with time, and to provide information under various pool conditions. In existing gravity dams and other structures with drainage systems, the most economical way of improving stability is usually to clean out existing drains or to add new foundation drains. The effectiveness of new drains however, will depend on the nature of the foundation, and upon whether or not the new drains intersect pressurized discontinuities. For structures without drainage systems, the uplift pressures can be reduced through the use of impervious blankets and cutoffs, although these methods of uplift reduction are much less reliable. Ongoing monitoring is required for drains, impervious blankets, and cutoff systems to ensure actual uplift pressures are within the range of values assumed for stability evaluation.

b. Structural anchor systems. Tensioned vertical or inclined anchors can be installed to improve sliding stability. Vertical anchors installed across potential failure planes improve the sliding resistance by adding a normal force that increases the shear-friction component of the sliding shear resistance. Inclining the anchors in the upstream direction will provide a horizontal component that can be used for additional shear resistance. However, untensioned anchors should not be used to improve sliding or rotational stability. Anchor systems are covered in detail in Chapter 8.

c. Adding concrete and buttressing to improve stability. Adding weight to a structure will increase compressive stress on failure planes and, thereby, increase the shear-frictional resistance to sliding. Concrete can be added to the front face, back face, or on top of the structure to increase the weight. The structure can also be buttressed with concrete blocks or with soil to improve sliding resistance. Often concrete blocks are placed against the toe of the structure and anchored with stressed tendons or bars (tensioned structural anchors) to further increase sliding resistance.

7-4. Case Histories

a. Eisenhower and Snell locks. Eisenhower and Snell locks are located on the St. Lawrence Seaway. The Corps of Engineers designed and constructed the two locks for the St. Lawrence Seaway Commission in the 1950s. Since that time, Corps personnel participated in annual inspections of the two locks. Cracking had been observed in the lock monoliths. Advanced analytical studies were required to determine the adequacy of the existing lock monoliths in terms of strength and stability, and to determine the advisability of lock monolith rehabilitation. A nonlinear finite element analysis, where loads are applied incrementally to simulate the actual field conditions, was used to evaluate monolith performance. Six categories were modeled in the FEM analysis:

- Initial conditions after excavation.
- Placement of the lock wall.
- Raising of the water table in the backfill.
- Formation of monolith cracks.
- Placement of tensioned structural anchors within the monoliths.
- Operation of the lock after rehabilitation.

The refined analysis procedure demonstrated the lock monoliths, once rehabilitated by installing tensioned structural anchors within the monoliths, could operate safely without structural or stability problems. This could not have been accomplished using traditional stability analysis procedures. (Mosher, Bevins, and Neeley 1991).

b. Lock 27 study. Lock 27 is located on the Mississippi River. Monolith 7E is a gravity earth-retaining structure that when evaluated using conventional equilibrium-based analysis methods demonstrated there was crack propagation at the structure-foundation interface. The crack propagation was evaluated using a finite element analysis procedure, a fracture-mechanics discrete crack analysis procedure, and a fracture-mechanics smeared crack analysis procedure. The finite element analysis utilized the SOILSTRUCT- Alpha procedure described in Chapter 6 above. The special-purpose FEM code MERLIN was used for the discrete crack analysis procedure, and the special-purpose FEM code CG-DAMS was used for the smeared crack analysis procedure. The crack estimated propagation length at the base of Monolith 7E was similar in each of the three advanced analytical methods and significantly less than that predicted by the conventional equilibrium-based analysis method. Base pressure distribution was also significantly different from the linear distribution assumed for the traditional method of analysis (Ebeling 1996).

c. Stewart Mountain Dam. Stewart Mountain Dam is a thin arch dam built in 1929 to 1930. An evaluation of the dam revealed that it was not dynamically stable during a Maximum Credible Earthquake (MCE) event; therefore, to stabilize the dam, 62 tensioned, structural anchors on 2.6m (8.5ft) to 3.0m (10.0ft) center-to-center spacing were installed from the crest of the dam and generally extended into the foundation. The bearing plates were designed to withstand a 4,450kN (1,000k)-tension load. A new overlay of 35mP (5,000psi) compressive strength concrete was placed on the crest of the dam to withstand punching shear. This overlay was designed to act as a beam on a flexible foundation because of the differences in the modulus of elasticity of the new overlay concrete and that of the existing dam concrete. A two-dimensional ADINA finite-element study was performed on a typical localized cross section of the top of the dam. The bearing plate, overlay, and existing dam concrete were modeled. Reinforcing for the overlay was designed to withstand the computed horizontal tension stresses under the bearing plate and the shear stresses below the edges of the bearing plate. For a detailed discussion of the design, refer to Bureau of Reclamation 1991.

Chapter 8
Anchoring Structures

8-1. General

Structural anchors are often used to improve the stability of existing structures but, generally, should not be used as a primary means to stabilize new, large mass concrete structures. If a situation exists where anchors are used to stabilize a new structure (space limitations and/or economics dictate their use), prior approval from CECW-E must be obtained. This chapter provides guidance on the design of anchoring structures to rock, it does not include information on soil anchors. Use of soil anchors for permanent structures is not allowed without prior approval from CECW-E. Untensioned structural anchors are commonly used in designs of new structures to stabilize thin concrete members on or against competent rock. Thin concrete members include walls, chute slabs, stilling basin slabs, and paved channel slabs. For such uses, prior approval from CECW-E is not required. Following are brief descriptions of new projects, which did utilize anchors into rock to stabilize large structures.

a. Bay Springs Lock. This lock was located in a deep-rock excavation so a considerable savings was achieved by anchoring the lock walls to the rock. This reduced the amount of rock excavation and the amount of mass concrete. This project was designed by the Nashville District.

b. Bonneville Lock. This lock and the upstream approach channel wall are in close proximity to a railroad. So, to avoid relocating the railroad it was determined the most economical design was to use structural anchors to stabilize the walls. This project was designed by the Portland District.

c. Wanapum Dam. Tensioned structural anchors have also been used by others to stabilize new structures. The Wanapum project included a hydropower station designed to provide for six future generating units. To reduce initial costs, only the intakes for these six future generating units were constructed. Since the entire structure was not built, the intakes lacked stability. The most practical and economical solution was to use tensioned structural anchors to stabilize the monoliths (Eberhardt and Veltrop 1965).

8-2. Anchoring Structures to Rock

a. Tensioned structural anchors. Inclined, tensioned structural anchors are effective in increasing the stability of structures by improving factors of safety against sliding and uplift, and improving the resultant location. The number, orientation, and capacity of anchors should be based on engineering considerations and stability requirements. The process of installing tensioned anchors consists of:

- drilling holes across the potential failure plane to accommodate anchor installation
- installing the anchors
- grouting the fixed or dead end of the anchors
- jacking at the stressing or live end of the anchor to induce an active tensile force which will provide a clamping action across the potential failure plane
- grouting the remaining ungrouted length of the anchor for corrosion protection.

Fixed end anchorage is created by grouting the annular space around the tendons for a length of tendons (bond length) sufficient to develop the ultimate tensile capacity of the anchorage system. Full active tensile load is developed in the remaining ungrouted (free stressing length) of the tendon anchor system by jacking at the free end of the tendons. The tensile force induced in the free stressing length of the anchor loads the potential failure plane with a compressive stress normal to the failure plane and for inclined anchors, a horizontal component that counteracts applied shearing loads. Allowances are made in the jacking load for anticipated losses such as relaxation of the tendons, seating losses, creep, and foundation consolidation.

b. Anchor System Selection. There are a large number of products and systems to choose from, with the primary considerations being cost and corrosion protection. Generally, fewer higher capacity anchors are less expensive than more smaller capacity anchors, however with high capacity anchors, the failure of a single anchor can result in a significant loss of sliding resistance along the potential failure plane. Single, high-strength steel bars have the smallest capacity of tensioned anchors, while steel strand is used when large capacity anchors are needed. Single bars provide less force because they have a lower working stress than strands. The three most common types of anchor systems are the hollow groutable anchors, the solid continuously threaded bar anchors, and strand anchors. Following is a brief description of each system.

(1) The hollow-groutable anchor system consists of a threaded high-strength reinforcing bar with a hole in the middle to permit the annular space around the bar to be grouted after tensioning is complete. Anchorage of the dead end is accomplished by an anchor cone, which engages the existing concrete. The bar is tensioned by hydraulic jack, or by tightening the nut located at the stressing end of the unit.

(2) The threaded-bar anchor system is tensioned with a center pull jack. A special threaded nut is used for locking- in the applied tension. The threaded bars may be secured in two-stage cement grout or fast- and slow-set polyester resin grouts. Long-term creep of polyester resin has been a concern that stems from laboratory tests at high-bond stresses. Generally, creep at normal working-level stresses will not be a problem.

(3) The strand anchor system is a seven-wire type having a center wire wrapped tightly by six outer wires. Strand anchors use single- or two-stage grouting with the single-stage grouting being referred to as an unbonded anchor, while the two-stage grouting procedure is referred to as a bonded anchor. The free stressing length of bonded anchors are grouted after the anchor has been stressed.

c. Untensioned structural anchors. Untensioned anchors shall not be used to improve sliding stability or resultant location since they may require excessive displacement before becoming effective. However, untensioned structural anchors may be used for the following situations.

(1) Walls. Anchored walls should be considered in areas where the rock is strong and firmly in place but is subject to surface expansion and exfoliation upon exposure. The concrete wall acts as a lining, which serves to limit the surface deterioration and to protect it against erosion from flowing water. The lining should be anchored adequately by untensioned structural anchors grouted into drilled holes in the rock. Unless the rock mass is essentially free of jointing, the anchorage should extend deep enough to engage a sufficient mass of rock for stability with a factor of safety where the anchored portion is assumed to be separated from the adjacent rock by continuous cracks. In some cases, the existence of definite slip planes dipping into the channel may require deeper anchorage near the top of the wall. Anchors may be installed at any inclination, except inclinations less than 5 degrees from horizontal, which should be avoided due to difficulties with installation. Horizontal and upwards-sloping anchors require specialized grouting techniques. Drainage should be provided to reduce the hydrostatic pressure behind the wall as much as practicable. Drain holes drilled into rock with outlets through the concrete generally provide the best and simplest system. The drain holes are sloped down a little toward the outlet and should be deep enough to adequately drain those cracks within that rock mass necessary for stability, plus the area immediately behind the mass. In cold climates, concrete thickness required for frost protection of poor rock may determine the wall thickness. Otherwise, the anchorage and concrete lining are designed to withstand the reduced hydrostatic pressure.

(2) Slabs. Anchored slabs on rock are designed to withstand uplift and other probable forces with an ample factor of safety. If the slab is on suitable rock, the slab can be anchored with untensioned structural anchors grouted in holes drilled into the rock. The holes shall have a sufficient depth to engage a mass of rock the submerged weight of which will withstand the net upward forces, assuming the mass of rock bounded by a 90-deg apex angle at the bottom with allowance for overlap from adjacent anchors. Drain holes drilled into the rock and discharging through the slab can reduce unbalanced hydrostatic uplift. Depth and spacing of drain holes should be determined by consultation with the geotechnical engineer. The holes should be inclined at a small angle from normal to the slab so that their outlet end is downstream from their inlet end to avoid a possible increased uplift from flowing water. A pervious drainage blanket under the slab with transverse perforated drain pipes discharging through the slab may be economical where a blanket is needed to protect the foundation from freezing.

8-3. Tensioned Anchor Loads

For inclined structural anchors, the normal component of the anchor force adds to the frictional resistance by increasing the normal force on the sliding plane, and the tangential component acts against the driving forces as resisting forces. The horizontal component should be treated as an applied load (see example D5 in Appendix D). Adding the horizontal component to the frictional resistance is not logical since the frictional resistance is a function of the normal loading and angle of internal friction, while the tangential component is related only to tendon load and inclination of the tendon. These applied loads occur immediately upon stressing the anchors and are not associated with deformations accompanying the development of frictional resistance. Thus, it should be considered an applied load that induces shear stresses along potential sliding planes opposite in direction to those due to downstream forces on the structure.

Rehabilitation of existing structures should examine a range of anchor forces varying from the force needed to meet Chapter 3 stability requirements to minimal forces needed to meet the requirements discussed in Chapter 6. A check should be made to ensure that the tangential component is never so large as to cause upstream sliding. An example might be that the tangential component applied to a dam is equal to or larger than the reservoir load on a dam. This raises the question of possible upstream sliding should the reservoir ever be evacuated.

8-4. Structural Anchor Design

a. Tensioned structural anchors. Recommendations for Prestressed Rock and Soil Anchors (Post-Tensioning Institute 1996) provides guidance in the application of permanent and temporary utilizing high-strength prestressing steel (includes bars and strand guidance). The recommendations do not deal with the design of anchored structures in general, but are limited to considerations specific to the prestressed anchors. Design, materials, fabrication and handling, installation, and performance testing of permanent tensioned structural anchors should be in accordance with PTI 1996 subject to the following provisions.

- PTI Chapter 3.0, "Specifications and Responsibilities" is not applicable because the structural anchors are a part of a larger Corps project and, as such, the structural anchors are included as a part of the contract documents.
- PTI Chapter 4.0, "Materials" requirements shall be used for prestressing steel (strand, bar, indented strand, epoxy-coated strand, epoxy-coated bar), anchorages, couplers, centralizers and spacers, corrosion inhibiting compound, bond breaker, sheath, tendon bond-length encapsulations, heat-shrinkable sleeves, grout tubes, anchor grout, cement grout, and polyester resin grouts.
- PTI Chapter 5.0, "Corrosion Protection," provides guidance for two classes of corrosion protection. Class I, encapsulated tendons, is often referred to as double corrosion protection while Class II, grout protected tendons, is often referred to as single corrosion protection. Class I corrosion protection shall be used for the anchors.
- PTI Paragraph 6.4.2 discusses fully bonded and unbonded anchors. Fully bonded anchors are required and shall be grouted in two stages, i.e., first stage (primary) is to grout the bond zone, and the second stage is to grout the free length after the anchor has been stressed.
- PTI Paragraph 7.4 discusses waterproofing rock-anchor drill holes. Provisions in this paragraph are a requirement.

b. Untensioned structural anchors. There is no industry standard for the design of untensioned structural anchors. Such anchors shall be ASTM grade 60 bars designed in accordance with EM 1110-2-2104. Anchors are usually of large diameter to reduce drilling cost. With large anchors however, the strain penetration that occurs in the reinforcement on each side of the failure plane is large, requiring significant elongation to develop the ultimate capacity of the anchor.

c. Bond length for tensioned and untensioned structural anchors. The required anchorage length is based upon tendon load, diameter of the tendon hole, type of tendon system, and the bond strength between the surrounding material and grout and between the tendon and grout. The bond lengths calculated below should be verified by performing pull-out tests utilizing tendon systems and drill-hole sizes matching those used by the contractor. If the foundation is variable, pull-out tests should be performed in areas representative of all site conditions and bond lengths adjusted accordingly. The embedment, and/or hooks, in the concrete for untensioned structural anchors should be in accordance with the requirements of EM 1110-2-2104. Embedment length should not be confused with the anchorage bond length of the tendon. Bond lengths for tensioned and untensioned structural anchors can be determined by using the largest value resulting from the computations in paragraphs (1), (2), or (3) below.

(1) Bond between tendon and grout

$$L = P / (\varphi \pi d_b f_{bu})$$

where: L = bond length

P = ultimate load for untensioned anchors and design load for tensioned anchors

d_b = diameter of tendon

f_{bu} = ultimate bond strength between tendon and grout (For design, use f_{bu} equal to $6\sqrt{f_c'}$ where f_c' is the compressive strength of grout; f_{bu} must be verified by tests as specified in PTI 1996)

φ = strength reduction factor = 0.9 (Not to be confused with soil strength parameter.)

(2) Bond between grout and rock

$$L = P / (\varphi \pi d_h f_{cu})$$

Where: d_h = diameter of drilled hole

f_{cu} = ultimate bond strength between grout and rock, or c, whichever is less (see Table 6.1 of PTI 1996 for values of f_{cu})

c = cross bed shear value in rock in the direction of pull out

(3) Rock-mass shear failure

(a) Tensioned structural anchors. With all tensioned structural-anchor systems, a major consideration is determining how deep to install the anchors. An anchor system that is too shallow may cause tension and cracking to occur along potential failure planes in the foundation, and a system too deep is uneconomical. PTI recommends normal bond length not less than 3.0m (10ft) for bars and 4.5m (15ft) for strand. Bond lengths greater than 10m (35ft) are normally not used. PTI recommends free stressing lengths to be at least 3.0m (10ft) for bar tendons and 4.5m (15ft) for strand tendons. Center-to-center spacing between anchors shall be at least 1.5m (5ft) unless unusual circumstances dictate. The fixed end (dead end) anchorages should be staggered.

(b) Untensioned structural anchors. For untensioned structural anchors, rock-mass failure will not normally govern design. Where the foundation consists of weak or cracked rock, a rational analysis should be performed to ensure adequate embedment length. This analysis should be performed in accordance with the requirements given in EM 1110-2-2400.

d. Anchorages for tensioned structural anchors. Anchorages shall be a combination of either a steel bearing plate and wedges, or a steel-bearing plate with a threaded anchor nut. The anchorage components must be carefully designed, detailed and constructed to provide adequate corrosion protection. It is also important to properly detail the region under the anchorage. Large tendons will exert very large forces on the bearing plate, which will then be distributed into the concrete. This will result in localized stress concentrations in the concrete at the edges and/or immediately beneath the bearing plate. Anchor details and bearing plate requirements are available from the post-tensioning system manufacturer. Guide specifications for post-tensioning materials are available from the Post Tensioning Institute (PTI), and contain information on permissible compressive concrete stresses beneath stressing-end anchor plates.

8-5. Stressing, Load Testing, and Acceptance

Field tests shall be performed before and during the installation to verify the adequacy of the anchor system and installation procedures. Tests before installation shall be used to check the performance of selected drilling method, conformability of hole size and drift tolerances, adequacy of assumed bond strengths between grout and rock, grout and concrete, and grout and tendon. Tests during installation should be adequate to ensure that anchors are installed in accordance with the requirements of the plans and specifications and that the ultimate capacity of the tendon can be developed. The number of tests required depends on the site-specific information including drilling conditions, type and/or size of tendons, and complexity of foundation formation and material. All testing procedures and acceptance criteria shall be in accordance with PTI 1996.

8-6. Monitoring Structural Anchor Performance

Grouting, which is required to protect tensioned anchors against corrosion, results in a fully bonded anchor. Untensioned anchors are installed with a single grouting process, which also results in a fully bonded anchor. Therefore, the long-term monitoring of anchor systems is limited. It is assumed that fully bonded anchors are maintenance free because the alkalies in the grout provide extra corrosion protection, the entire anchor is not lost if a portion of the anchor fails, and small, localized movements can mobilize reserve capacity in the anchor. Inspection of the structure and monitoring of its behavior are the only practical ways of monitoring performance of fully bonded, structural anchors. Provisions are being developed that will allow measurements of galvanic action within the anchor system. Also, sensors can now be embedded in fiber-optic cables. These cables have been epoxied to ship hulls, bridge girders and reinforcing bars with the sensors transmitting data in real time. It is believed they could be epoxied to anchors. These procedures may be investigated by the designers if there is a need for further assurance that an anchor system remains functional.

Appendix A
References

A-1. Required Publications

ER 1105-2-101
Risk-Based Analysis for Evaluation of Hydrology/Hydraulics, Geotechnical Stability, and Economics in Flood Damage Reduction Studies

ER 1110-2-100
Periodic Inspection and Continuing Evaluation of Completed Civil Works Structures

ER 1110-2-1150
Engineering and Design for Civil Works Projects

ER 1110-2-1155
Dam Safety Assurance Program

ER 1110-2-1806
Earthquake Design and Evaluation of Civil Works Projects

EM 1110-1-1802
Geophysical Exploration for Engineering and Environmental Investigations

EM 1110-1-1804
Geotechnical Investigations

EM 1110-1-1904
Settlement Analysis

EM 1110-1-1905
Bearing Capacity of Soils

EM 1110-1-2908
Rock Foundations

EM 1110-2-1603
Hydraulic Design of Spillways

EM 1110-2-1612
Ice Engineering

EM 1110-2-1614
Design of Coastal Revetments, Seawalls, and Bulkheads

EM 1110-2-1619
Risk-Based Analysis for Flood Damage Reduction Studies

EM 1110-2-1901
Seepage Analysis and Control for Dams

EM 1110-2-1906
Laboratory Soils Testing

EM 1110-2-1913
Design and Construction of Levees

EM 1110-2-2007
Structural Design of Concrete Lined Flood Control Channels

EM 1110-2-2104
Strength Design for Reinforced-Concrete Hydraulic Structures

EM 1110-2-2200
Gravity Dam Design

EM 1110-2-2201
Arch Dam Design

EM 1110-2-2400
Structural Design of Spillways and Outlet Works

EM 1110-2-2502
Retaining and Flood Walls

EM 1110-2-2602
Planning and Design of Navigation Locks

EM 1110-2-2607
Navigation Dam Masonry

EM 1110-2-3001
Planning and Design of Hydroelectric Power Plant Structures

EM 1110-2-3104
Structural and Architectural Design of Pumping Stations

American Society of Civil Engineers 1990
American Society of Civil Engineers. 1990. Minimum Design Loads for Buildings and Other Structures, ASCE 7, NY.

American Society for Testing and Material 1990
American Society for Testing and Materials. 1990. Standard Test Method for Classification of Soils for Engineering Purposes, Practice No. D2487-90, 1990 Book of ASTM Standards, 04.08, 309-319.

Electric Power Research Institute (1992)
Electric Power Research Institute. 1992. Uplift pressures, shear strengths, and tensile strengths for stability analysis of concrete gravity dams, TR-100345, 3142 Hillview Avenue, Palo Alto, CA.

Post-Tensioning Institute 1996
Post-Tensioning Institute. 1996. Recommendations for Prestressed Rock and Soil Anchors, Phoenix, AZ.

Rock Testing Handbook 1980
Rock Testing Handbook, 1980.
U.S. Army Engineer Waterways Experiment Station, U.S. Government Printing Office, Washington, DC.

Shore Protection Manual 1984
Shore Protection Manual, 1984. 4th ed., vol 2,
U.S. Army Engineer Waterways Experiment Station, U.S. Government Printing Office, Washington, DC.

A2. Related Publications

Bazant 1990
Bazant, Z.P. 1990. A critical appraisal of no-tension dam design: a fracture mechanics viewpoint, *Dam Engineering*, 1(4).

Bureau of Reclamation 1991
Bureau of Reclamation. 1991. Structural design of post-tensioned cables for the dynamic stability of Stewart Mountain Dam, TM No. SMC-3110-01-90, Phoenix, AZ.

Casagrande 1937
Casagrande, A. 1937. Seepage through dams. *Contributions to soil mechanics, 1925-1949*, Boston Society of Civil Engineers, Boston, MA, 295-336.

Casagrande 1973
Casagrande, Leo. 1973. Comments of conventional design of retaining structures, *Journal, Soil Mechanics and Foundations Division, American Society of Civil Engineers* 99(SM 2), 181-197.

Cedergren 1967
Cedergren, Harry R. 1967. Seepage, drainage, and flow nets. John Wiley & Sons, NY

Clough and Duncan 1969
Clough, G. W., and Duncan, J. M. 1969. Finite element analyses of Port Allen and Old River Locks, Contract Report S-69-6, U.S. Army Engineer Waterways Experiment Station, Vicksburg, MS.

Dewey, Reich, and Saouma 1994
Dewey, R. R., Reich, R. W., and Saouma, V. E. 1994. Uplift modeling for fracture mechanics analysis of concrete dams, *ASCE Journal of Structural Engineering*, 120 (10).

Duncan, Clough, and Ebeling 1990
Duncan, J. M., Clough, G. W., and Ebeling, R. M. 1990. Behavior and design of gravity earth retaining structures, invited contribution to the ASCE conference *Design and Performance of Earth Retaining Structures,* Cornell, NY.

Ebeling 1990
Ebeling, R. M. December, 1990. Review of finite element procedures for earth retaining structures, Miscellaneous Paper ITL-90-5, Information Technology Laboratory, U.S. Army Engineer Waterways Experiment Station, Vicksburg, MS.

Ebeling 1996
Ebeling, R. M. 1996. Locks 27 gravity retaining wall, *The REMR Bulletin*. U.S. Army Engineer Waterways Experiment Station, Vicksburg, MS.

Ebeling, et al. 1992
Ebeling, R. M., Clough, G.W., Duncan, J.M., and Brandon, T.L. 1992. Methods of evaluating the stability and safety of gravity earth-retaining structures founded on rock, Technical Report REMR-CS-29, U.S. Army Engineer Waterways Experiment Station, Vicksburg, MS.

Ebeling, Duncan, and Clough 1990
Ebeling, R. M., Duncan, J. M., and Clough, G. W. 1990. Methods of evaluating the stability and safety of gravity earth-retaining structures founded on rock, Phase 2 Study, Technical Report ITL-90-7, U.S. Army Engineer Waterways Experiment Station, Vicksburg, MS.

Ebeling, Mosher, Abraham, and Peters 1993

Ebeling, R. M., Mosher, R. L., Abraham, K., and Peters, J. E., 1993. Soil-structure interaction study of Red River Lock and Dam No. 1 subjected to sediment loading, Technical Report ITL-93-3. U.S. Army Engineer Waterways Experiment Station, Vicksburg, MS.

Ebeling, Clough, Duncan, and Brandon 1992
Ebeling, R. M., Clough, G. W., Duncan, J. M., and Brandon, T. .L. 1992. Methods of evaluating the stability and safety of gravity earth-retaining structures founded on rock, Technical Report REMR CS-29, U.S. Army Engineer Waterways Experiment Station, Vicksburg, MS.

Ebeling, Morrison, and Mosher 1996
Ebeling, R. M., Morrison, E. E., and Mosher, R. L. 1996. Stability of existing concrete structures, *The REMR Bulletin,* 13(1). U.S. Army Engineer Waterways Experiment Station, Vicksburg, MS.

Ebeling, and Mosher 1996
Ebeling, R. M., and Mosher, R. L. 1996. Red River U-Frame Lock No.1, Backfill-Structure-Foundation Interaction, *ASCE Journal of Geotechnical Engineering* 122(3), pp 216-225.

Ebeling, et al. 1993
Ebeling, R. M., Mosher, R. M., Abraham, K., and Peters, J. F. 1993. Soil-structure interaction study of Red River Lock and Dam No. 1 subjected to sediment loading, Technical Report ITL-93-3, U.S. Army Engineer Waterways Experiment Station, Vicksburg, MS.

Ebeling, Peters, and Clough 1992
Ebeling, R. M., Peters, J., and Clough, G. W. 1992. Users guide for the incremental construction, soil-structure interaction program SOIL-STRUCT, Technical Report ITL-90-6, U.S. Army Engineer Waterways Experiment Station, Vicksburg, MS.

Ebeling and Morrison (1992)
Ebeling, R. M. and Morrison, E.E., November 1992, Seismic design of waterfront retaining structures, Technical Report ITL-92-11, US Army Engineers Waterways Experiment Station, Vicksburg, MS

Ebeling and Morrison (1997)
Ebeling, R. M. and Morrison, E.E., July 1997, Evaluating the stability of existing massive concrete gravity structures founded on rock, Technical Report ITL-97-5, US Army Engineers Waterways Experiment Station, Vicksburg, MS

Ebeling, Pace and Morrison (1992)
Ebeling, R. M., Pace. M.E., and Morrison, E.E., 1992, The seismic design of waterfront retaining structures, Technical Report ITL-92-11, US Army Engineers Waterways Experiment Station, Vicksburg, MS

Ebeling, Peters and Mosher (1997)
Ebeling, R.M., Peters, J.F., and Mosher 1997 "The Role of Non-Linear Deformation Analyses in the Design of a Reinforced Soil Berm at Red River U-Frame Lock No. 1," International Journal for Numerical and Analytical Methods in Geomechanics, Vol. 21, pp. 753-787.

Ebeling and Wahl (1997)
Ebeling, R. M and Wahl, R. E., Soil-structure-foundation iteration analysis of the new roller-compacted concrete north lock wall at McAlpine Lock, Technical Report REMR-CS-54, July 1997, US Army Engineers Waterways Experiment Station, Vicksburg, MS

Ebeling and Filz (1998)
Ebeling, R.M.and Filz, G.M., 1998. Soil-structure interaction analysis of rock-founded gravity and cantilevered retaining walls, Technical Report in publication, U.S. Army Engineer Waterways Experiment Station, Vicksburg, MS.

Eberhardt and Veltrop (1965)
Eberhardt, A., and Veltrop, J. A. 1965. 1300 ton capacity prestressed anchors stabilize dam, *Journal of the Prestressed Concrete Institute (PCI)*, 10(4).

Filz, and Duncan, (1992)
Filz, G. M., and Duncan, J. M., 1992, An experimental and analytic study of earth loads on rigid retaining walls, Geotech. Engrg. Res. Rep., Virginia Polytech. Inst. And State University, Blacksburg, Va.

Filz, Duncan, and Ebeling (1997)
Filz, G. M., Duncan, J. M., and Ebeling, R. M., September 1997. Vertical shear forces on non-moving walls: applications, *American Society of Civil Engineers Geotechnical Journal* 123(9)

Filz and Duncan 1992
Filz, G. M., and Duncan, J. M. 1992. An experimental and analytic study of earth loads on rigid retaining walls, Technical Report to the U.S. Army Engineer Waterways Experiment Station, Virginia Polytechnic Institute and State University, Blacksburg, VA.

Harr 1962
Harr, M. E. 1962. *Groundwater and seepage*. McGraw-Hill, NY.

Hilmer 1986
Hilmer, K. 1986. Evaluation of a ten-year measuring program at the Eibach Lock.

Lysmer, et al. 1975
Lysmer, J., Utaka, T., Tsai, C-H, and Seed, H.B. 1975. "FLUSH, A Computer Program for Approximate 3-D Analysis of Soil-Structure Interaction Problems," Earthquake Engineering Research Center, Report No. EERC 75-30, College of Engineering, University of California, Berkeley, CA.

Matsuo, Kenmochi, and Yagi 1978
Matsuo, M., Kenmochi, S., and Yagi, H. 1978. Experimental study on earth pressure of retaining wall by field tests, *Soils and Foundations*, Japanese Society of Soil Mechanics and Foundation Engineering, 18(3). Japanese Society of Soil Mechanics and Foundation Engineering, Suga-yama Building 4F, Kanda Awaji-cho 2-23 Chiyoda-ku, Tokyo, Japan.

Mononobe and Matsou 1929.
Mononobe, N., and Matsuo, H. 1929, On the determination of earth pressures during earthquakes, Proceedings, World Engineering Congress, 9.

Okabe 1926
Okabe, S. 1926. "General Theory of Earth Pressures," Journal Japan Society of Civil Engineering, Vol. 12, No. 1.

Saouma, Bruhwiler, and Boggs 1990
Saouma, V.E., Bruhwiler, E., and Boggs, H.L. 1990. A review of fracture mechanics applied to concrete dams, *Dam Engineering* 1(1).

Seed and Whitman 1970
Seed, H. B., and Whitman, R. V. 1970. Design of earth retaining structures for dynamic loads, *ASCE Specialty Conference, Lateral Stresses in the Ground and Design of Earth Retaining Structures*, 103-147.

Terzaghi 1934
Terzaghi, K. 1934. Large retaining wall tests, *Engineering News - Record*, 85, February 1--April 19.

Tschebatarioff 1949
Tschebatarioff, G. 1949. Large scale earth pressure tests with model flexible bulkheads, Final Report to U. S. Navy, Bureau of Yards and Docks, Princeton University, Princeton, NJ.

Westergaard 1933
Westergaard, H. M. 1933. Water pressures on dams during earthquakes, *American Society of Civil Engineers*, 98, 418-433. ASCE Publication Fulfillment, 345 East 4th St., New York, NY.

Whitman and Liao 1985
Whitman, R. V., and Liao, S. 1985. Seismic design of gravity retaining walls, Miscellaneous Paper GL-85-1, U. S. Army Engineer Waterways Experiment Station, Vicksburg, MS.

Appendix B
Loading Conditions and Loading-Condition Classification

B-1. Purpose

The main purposes of Appendix B are to:

- Assemble, for convenience and easy reference, all loading-condition requirements for each structure type as specified in various Engineer Manuals.
- Update the classification of each loading condition to comply with the Table 3-1 definition for *usual, unusual,* and *extreme* (U/UN/E).
- Update older guidance to conform with current earthquake and flood loading definitions.
- Provide a U/UN/E loading classification designation for those loading conditions that may fall into either the usual, unusual, or extreme categories depending on the probability of occurrence.

B-2. Loading Conditions, Normal Operating Pool Condition and Coincident Pool Condition

a. General. Although the criteria in this manual is for stability analysis only, the following tables of loading conditions and loading condition classifications are applicable to both the stability evaluation and to the structural design of Corps projects. This is done so the designer can cross-reference between this EM and other guidance. The loading-condition identification number and description identified in this manual, unless otherwise indicated, is the same as that in the structure- specific guidance. Some of the loading conditions will apply to stability evaluation only, some to structural design/evaluation only, and others will apply to both stability and structural design/evaluation. The loading conditions in these tables should be sufficient to cover most situations. However, the structural engineer should closely examine each loading condition to ensure that it is applicable to the particular structure and site conditions, that the assigned classification of usual, unusual, or extreme is correct, and that there are no additional loading conditions which should be investigated. The structural engineer should also consider any unique or special loadings that may occur during construction. Loading condition tables for each structure type are provided below, followed by a description of those loading conditions.

b. Normal operating pool. The normal operating pool condition refers to water surface elevations, which represent maximum loading with a 10-year return period. For flood control projects this will correspond to 10-year maximum inflow conditions. For some navigation projects it may correspond to low flow conditions, which result in 10-year minimum tailwater, coincident with a fairly constant upper pool.

c. Coincident pool. The coincident pool represents temporal average pool conditions, which are used for load combinations that include seismic loads. The pool elevation that is equaled or exceeded 50-percent of the time is the temporal average pool. An estimate of temporal average headwater and tailwater pools can be obtained based on existing project operations data (providing sufficient data exists), or by using inflow data in combination with planned project operating procedures. A plot of this information is shown in Figure B-1. This figure was developed under the assumption the project is operated for a year under mean annual pool conditions, with project inflows varying from month to month due to seasonal variations. The mean monthly pool elevation from Figure B-1 can then be used to develop a pool-duration plot showing the percent of time a particular pool elevation will be equaled or exceeded. The pool elevation that is equaled or exceeded 50-percent of the time (or 182 days per year) is the temporal average pool. This is illustrated by Figure B-2.

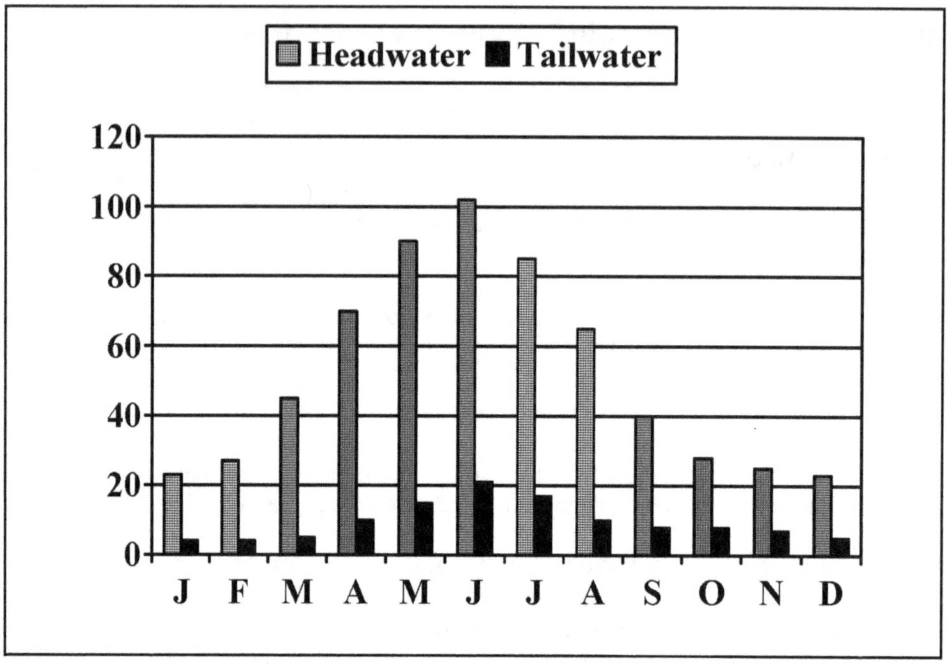

Figure B-1 Monthly Plot of Average Annual Water Levels

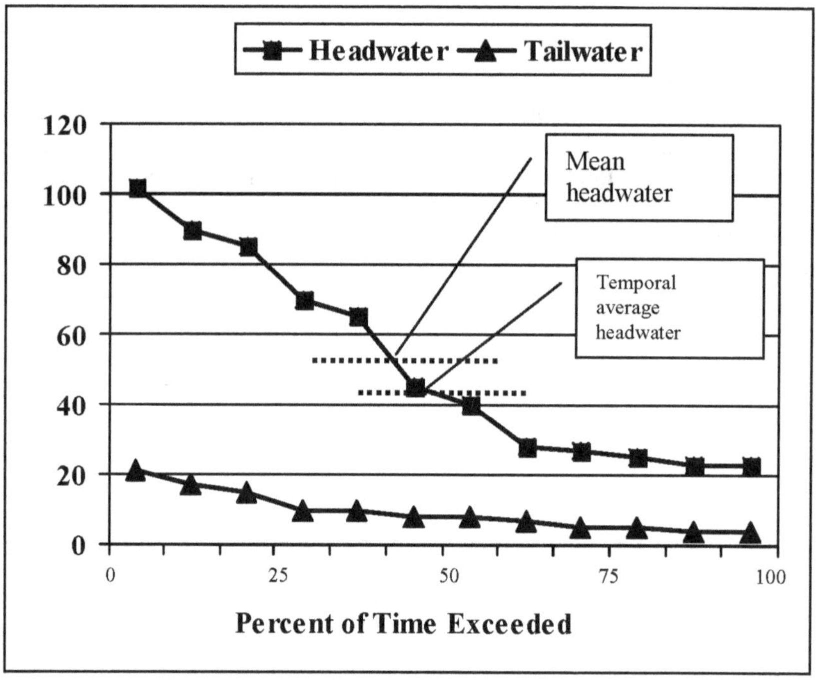

Figure B-2 Percent of Time Average Annual Pools are Equaled or Exceeded

B-3. Classification Tables

Table B-1 Gravity-Dam Loading-Condition Classification
Structure Type: Gravity Dams, EM 1110-2-2200; Navigation Dams, EM 1110-2-2607

Load Case	Loading Description	Classification*
1	Construction Condition	UN
2	Normal Operating	U
3	Infrequent Flood	UN
4	Construction with Operational Basis Earthquake (OBE)	E
5	Coincident Pool with OBE	UN
6	Coincident Pool with Maximum Design Earthquake (MDE)	E
7	Maximum Design Flood (MDF)	U/UN/E

* U = usual, UN = unusual, E = extreme

a. Gravity dam - basic loading conditions. The basic load cases of Table B-1 are generally used in the stability evaluation of concrete gravity dams. These load cases are described below. Load cases used in the stability analysis of powerhouses and power intake sections are covered in Table B-16. For the normal operating load condition, the normal operating pool is defined by Paragraph B-2b(1). For earthquake load conditions, the coincident pool is defined by Paragraph B-2b(2).

(1) Loading condition 1 - Construction.

- Dam structure complete.
- No headwater or tailwater.

(2) Loading condition 2 - Normal Operating.

- Headwater at normal pool as defined by Paragraph B-2b(1)
- Minimum tailwater corresponding with the above headwater.
- Uplift.
- Ice and silt pressure, if applicable.

(3) Loading condition 3 - Infrequent Flood.

- Pool at an elevation representing a flood event with a 300-year return period.
- Minimum corresponding tailwater.
- Uplift.
- Ice and silt pressure, if applicable.

(4) Loading condition 4 - Construction with OBE.

- OBE.
- Horizontal acceleration in upstream direction.
- No headwater or tailwater.

(5) Loading condition 5 - Coincident Pool + OBE.

- OBE, horizontal acceleration in downstream direction.
- Coincident pool condition as defined in Paragraph B-2b(2) with corresponding tailwater.
- Uplift at pre-earthquake level.

- Silt pressure, if applicable.
- No ice pressure.

(6) Loading condition 6 - Coincident Pool + MDE.

- MDE, horizontal acceleration in downstream direction.
- Coincident pool condition as defined in Paragraph B-2b(2) with corresponding tailwater.
- Uplift at pre-earthquake level.
- Silt pressure, if applicable.
- No ice pressure.

(7) Loading condition 7 - MDF.

- Combination of pool and tailwater, which produces the worst structural loading condition, with an unlimited return period. (See paragraph 4-2e for additional information on the MDF)
- Uplift.
- Silt, if applicable.
- No ice pressure.

Table B-2 Arch Dam Loading-Condition Classification
Structure Type: Arch Dams, EM 1110-2-2201

Load Case	Loading Description	Classification
SU1	Minimum-usual concrete temperature. Reservoir elevation occurring at that time. Dead Load.	U
SU2	Maximum-usual concrete temperature. Reservoir elevation occurring at that time. Dead Load.	U
SU3	Normal-operating pool condition. Concrete temperature occurring at that time. Dead Load.	U
SUN1	Infrequent Flood. Reservoir at elevation representing a 300 year flood event. Concrete temperature at that time. Dead Load.	UN
SUN2	Minimum-design reservoir elevation. Concrete temperature occurring at that time. Dead Load.	UN
SUN3	End of construction condition. Structure completed, empty reservoir. Temperature Load.	UN
SE1	Reservoir at MDF elevation. Concrete temperature occurring at that time. Dead load.	UN/E
DUN1	OBE plus coincident pool	UN
DUN2	Operating Basis Earthquake (OBE) plus static load case SUN3.	UN
DE1	Maximum Design Earthquake (MDE) plus coincident pool	E

b. Arch dams - basic loading conditions. Table B-2 describes the static and dynamic loading combinations to be used for evaluating sliding stability of arch dam abutments. These loading combinations are also used to evaluate stresses within the dam. The load cases covered in Table B-2 are similar to those in EM 1110-2-2201, and they should be sufficient to cover most arch dams; however, each loading condition should be carefully examined to ensure that it is applicable, and that it is properly classified under one of the three categories, i. e., usual, unusual, or extreme. The loading combinations should be established at the earliest stages of design and adhered to throughout the development of the final design. Since there are different factors of safety for different loading conditions, the selection and classification of the load cases greatly influence the geometry of an arch dam and the resulting stresses The structural engineer evaluating stability should refer to paragraph B-2b for the definition of Anormal operating@ and Acoincident@ pool conditions, and should refer to paragraphs 4-2 c and 4-2 e for the definitions of Infrequent Flood (IF) and Maximum Design Flood (MDF).

Table B-3 Retaining-Wall Loading-Condition Classification
Structure Type: Retaining Walls, EM 1110-2-2502

Load Case	Loading Description	Classification
R1	Normal Operating	U
R2	Normal Operating + Short Duration Loads	UN
R3a	Normal Operating + OBE	UN
R3b	Normal Operating + MDE	E

Table B-4 Inland Flood-Wall Loading-Condition Classification
Structure Type: Inland Flood Walls, EM 1110-2-2502

Load Case	Loading Description	Classification
I1	Infrequent Flood	UN
I2	MDF	UN/E
I3a	Coincident Pool + OBE	UN
I3b	Coincident Pool + MCE	E
I4	Construction	UN

Table B-5 Coastal Flood-Wall Loading-Condition Classification
Structure Type: Coastal Flood Wall, EM 1110-2-2502

Load Case	Loading Description	Classification
C1	Surge Stillwater + Coincident Wave	UN/E
C2a	Coincident Pool + OBE	UN
C2b	Coincident Pool + MDE	E
C3	Construction	UN
C4	Normal Operating	UN

c. Walls - basic loading conditions. Loading conditions for the various types of walls are summarized in Tables B-3 through B-5. These loading conditions are generally representative of conditions affecting retaining walls, inland flood walls, and coastal flood walls. The loading conditions in the tables are described in further detail in the following paragraphs. Note that some walls may require consideration of loadings from more than one of the above tables in cases where the wall will act as a retaining wall for one loading condition, and as a flood wall for another loading condition.

(1) Retaining walls.

(a) Loading Condition R1 - Normal Operating.

- Backfill is placed to the final elevation (the backfill is dry, moist, or partially saturated as the case may be).

- Surcharge loading, if present, is applied (stability should be checked with and without surcharge).
- Any existing lateral and uplift pressures due to water are applied.
- Construction loads, which are not considered short-duration loads.

(b) Loading Condition R2 - Normal Operating + Short Duration Loads. This case is the same as R1 except the water table level in the backfill rises, for a short duration, or another type of loading of short duration is applied.

(c) Loading Condition R3a - Normal Operating + OBE. This is the same as Case R1 except with the addition of OBE induced lateral and vertical loads. The uplift is the same as for Case R1.

(d) Loading Condition R3b - Normal Operating + MDE. This is the same as Case R1 except with the addition of Maximum Design Earthquake (MDE) induced lateral and vertical loads. The uplift is the same as for Case R1.

(2) Inland floodwalls.

(a) Loading Condition I1 - Infrequent Flood.

- Backfill in place to final elevation.
- Water at an elevation representing a flood with a 300-year return period (To be investigated only when the MDF is an extreme load condition). See paragraphs 4-2c and 4-2e for information regarding the IF and MDF.
- Uplift.

(b) Loading Condition I2 - MDF.

- Combination of water on the protected and unprotected side, which produces the worst structural loading condition, with an unlimited return period.
- Uplift.

(c) Loading Condition I3a - Coincident Pool + OBE. (Note: This load case need only be considered if the wall has a significant loading during the non-flood stage)

- Backfill in place to final elevation.
- Water, if applicable, is at an elevation that is coincident with mean annual non-flood operating conditions.
- Uplift, if applicable.
- OBE-induced lateral and vertical loads.

(d) Loading Condition I3b +MDE - Coincident Pool + MDE. (Note: This load case need only be considered if the wall has a significant loading during the non-flood stage)

- Backfill in place to final elevation.
- Water, if applicable, is at an elevation that is coincident with mean annual non-flood operating conditions.
- Uplift, if applicable.
- MDE-induced lateral and vertical loads.

(e) Loading Condition I4 - Construction Condition.

- Floodwall is in place with the loads added which are possible during the construction period but are of short duration.

(3) Coastal flood walls.

(a). Loading Condition C1 - Design Surge Stillwater + Design Coincident Wave.

- Design surge stillwater condition + the governing nonbreaking, breaking, or broken wave conditions coincident with the design surge stillwater condition.
- Uplift is acting, based on the surge stillwater.

(b) Loading Condition C2a - Coincident Pool+ OBE.

- Water at level representing mean annual tide pool conditions
- Uplift, if applicable, is acting.
- OBE induced lateral and vertical loads, if applicable, are acting.

(c) Loading Condition C2b - Coincident Pool + MDE.

- Same as condition C2a, except use MDE.

(d) Loading Condition C3 - Construction.

- Floodwall is in place with the loads added which are possible during the construction period but are of short duration.

(e) Loading Condition C4 - Normal Operating

- Water is at the highest level with a 10-year return period on the unprotected side.
- Uplift is acting

Table B-6 Intake Tower Loading-Condition Classification
Structure Type: Intake Towers, EM 1110-2-2400

Load Case	Loading Description	Classification*
U1	Normal Pool, All Gates Open	U
U2	Normal Pool, One or more Gates Closed	U
U3	Normal Pool, All Gates Closed	U
U4	Normal Pool with Silt	U
U5	Minimum Pool	U
UN1	Infrequent Flood, All Gates Open	UN
UN2	Infrequent Flood, One or more Gates Closed	UN
UN3	Infrequent Flood, All Gates Closed	UN
UN4	Construction	UN
UN5	Diversion	UN
UN7	Maintenance Bulkheads in Place	UN
UN8	OBE + Coincident Pool	UN
E1	MDE + Coincident Pool	E
E2	MDF	UN/E

d. Intake towers - basic loading conditions. Specific operational and site conditions from construction through project life and structure configuration may require that the stability loading conditions be modified, or that additional analysis of conditions be made. The loading conditions of Table B-6 are described below.

(a) Loading Condition U1 - Normal Pool, All Gates Open.

- Dead load of structure.
- Reservoir at normal pool as defined by Paragraph B-2b(1).
- Earth load (if any).
- Ice loads, if applicable.
- Uplift.
- Water surface inside structure drawn down to hydraulic gradient with all gates fully opened.
- Wave loads, if applicable.

(b) Loading Condition Case No.U2 - Normal Pool, One or More Gates Closed.

- Dead load of structure.
- Reservoir at normal pool as defined by Paragraph B-2b(1).
- One or more gates closed with other gates fully opened and water surface drawn down to hydraulic gradient in remainder of structure in combinations that produce the most unstable conditions.
- Earth load (if any).
- Ice loads, if applicable.
- Uplift.
- Wet-well full of water upstream from closed gate.
- Wave loads, if applicable.

(c) Loading Condition U3 - Normal Pool, All Gates Closed.

- Dead load of structure.
- Reservoir at normal pool as defined by Paragraph B-2b(1).
- Earth load (if any).
- Uplift.
- Wave loads, if applicable.

(d) Loading Condition U4 - Normal Pool with Silt.

- Reservoir with silt for the most critical of proceeding conditions U1 through U3.

(e) Loading Condition U5 - Minimum Pool.

- Reservoir empty or at minimum pool.
- Dead load of structure.
- Earth load (if any).
- Ice loads, if applicable.
- Wind load in the direction that would produce the most severe foundation pressures.
- Uplift.
- Wave loads, if applicable.

(f) Loading Conditions UN1 through UN3.

- Loading conditions UN1 through UN3 are the same as U1 through U3 except the reservoir, rather than at the normal pool condition, is at the infrequent flood stage meaning at an elevation representing a flood event with a 300-year return period.

(g) Loading Condition UN4 - Construction.

- Reservoir empty.
- Dead load of structure (partially or fully completed).
- Earth load (if any).
- Heavy construction equipment required on or near the structure during construction.
- Wind load in the direction that would produce the most severe foundation pressures.

(h) Loading Condition UN5 - Diversion.

- Reservoir at maximum elevation expected during diversion.
- Dead load of structure at diversion level completion.
- Earth load (if any).
- Heavy construction equipment required on or near the structure.

Wind load in the direction that would produce the most severe foundation pressures.

(i) Loading Condition UN7 - Maintenance Bulkheads in Place.

- Bulkheads in place, no water in structure downstream of bulkheads.
- Dead load of structure.
- Reservoir at maximum pool level at which bulkheads are used.
- Earth loads (if any).
- Uplift.

(j) Loading Condition UN8 - Coincident Pool + OBE.

- OBE for the most critical of the conditions U1 through U5 with the reservoir at the coincident pool elevation.
- No ice.

(k) Loading Condition E1 - Coincident Pool + MDE.

- MDE for the most critical of the conditions U1 through U5 with the reservoir at the coincident pool elevation.
- No ice.

(l) Loading Condition E2 - MDF.

- Pool at PMF elevation.
- All gates opened or closed, depending on project operating criteria.

Table B-7 Navigation Lock Loading-Condition Classification
Structure Type: Navigation Lock Walls, EM 1110-2-2602

Load Case	Loading Description	Classification
1A	Normal Operating Lower Pool in Landward Lock Chamber (Upper Pool in Riverward Lock Chamber)[1]	U
1B	Normal Operating Upper Pool in Landward Lock Chamber (Lower Pool in Riverward Lock Chamber)[1]	U
2A	Unusual Operating Same as 1A or 1B with Extreme Low Water	UN
2B	Maintenance Condition	UN
2C	1A and 1B with OBE	UN
2D	1A and 1B with MDE	E
3	Construction Condition	UN

Note [1]: Normal operating loading conditions in parentheses apply only if the lock has two adjacent chambers.

Table B-8 Navigation Lock Loading-Condition Classification
Structure Type: Navigation Lock – Upper and Lower Gate Bays, EM 1110-2-2602

Load Case	Loading Description	Classification
1A	Normal Operating, Gates Loaded	U
1B	Normal Operating, Gates Unloaded	U
2B	1A or 1B with Extreme Low Water	UN
2C & 2D	1A & 1B with Ineffective Drains	UN
2E	Maintenance Condition	UN
2A & 2F	1A & 1B with OBE	UN
2G & 2H	1A and 1B with MDE	E
3	Construction Condition	UN

Table B-9 Navigation Lock Loading-Condition Classification
Structure Type: Navigation Lock – Approach Walls, EM 1110-2-2602

Load Case	Loading Description	Classification
1A	Normal Operating + Barge Impact (T_r = 10 years)	U
1B	Normal Operating + Barge Impact (T_r = 300 years)	UN
1C	Normal Operating + Barge Impact (T_r = 1000 years)	E
2A	Normal Operating + OBE	UN
2B	Normal Operating + MCE	E
3	Construction Condition	UN

Table B-10 Navigation Lock Loading-Condition Classification
Structure Type: Navigation Lock – Upper and Lower Sills, EM 1110-2-2602

Load Case	Loading Description	Classification
1A	Normal Operating	U
2A	1A with Extreme Low Tailwater	UN
2B	Maintenance Condition	UN

e. Navigation locks - basic loading conditions. Navigation locks may have land walls, river walls, intermediate walls, upper- and lower-gate bays, upper- and lower-approach walls , and upper- and lower-sill structures (Tables B-7 through B-10). Different operating and flood- discharge conditions can result in maximum head differentials and uplift pressures for each of these structures. Maintenance conditions, where the lock chambers and filling systems are unwatered, can sometimes govern the design. Loss of pool and rapid drawdown conditions, if possible, must also be evaluated. The effectiveness of drains used to reduce the water table in backfill materials and in the foundation must be considered in the stability analyses. Hawser and barge impact loads must be considered in the stability of navigation lock structures. Hawser and barge impact loads when combined with other normal-operating-condition loads are considered to be an unusual load condition. After the amount and intensity of the individual loads acting on the lock walls or monoliths have been determined as separate considerations, the possible combinations of such loads must be examined to determine the most adverse condition. Loadings that will ordinarily require examination are described in the following paragraphs. However, independent checks of each structure under consideration should be made to determine whether these conditions are adequate for determination of the most critical loading.

(1) Lock chamber walls (including land wall, river wall, and intermediate wall)

(a) Loading Condition 1A - Normal Operating Condition.

- Backfill loads (soil, water, and surcharge).
- Hawser load.
- Lower pool in landward lock chamber, upper pool in riverward lock chamber.
- Uplift as defined by water elevations.
- Vessel impact (Mean annual impact force).

(b) Loading Condition 1B - Normal Operating Condition.

- Loads (soil, water, and surcharge). Backfill.
- Hawser load.
- Upper pool in landward lock chamber, lower pool in riverward lock chamber.
- Uplift as defined by water elevations.
- Vessel impact

(c) Loading Condition 2A - Operating Condition With Drawdown. The same requirements for Conditions 1A and 1B are included except for the following conditions.

- Extreme low water stage for lower pool.

(d) Loading Condition 2B - Maintenance Condition. The same requirements for Conditions 1A and 1B are included except for the following conditions.

- Lock chamber unwatered to a predetermined level.
- No hawser load.

(e) Loading Condition 2C - Normal Operating +OBE. The same requirements for Conditions 1A and 1B except for the condition of an OBE earthquake load added in the most critical direction.

(f) Loading Condition 2D - Normal Operating + MDE. The same requirements for Conditions 1A and 1B except for the condition of an MDE earthquake load added in the most critical direction.

(g) Loading Condition 3 - Construction Conditions.

- Backfill loads (soil and surcharge).
- Wind as applicable.
- No uplift.
- Hydrostatic forces are active in accordance with construction or cofferdam plans.

(2) Upper and lower gate bays.

(a) Loading Condition 1A - Normal Operating Condition.

- Upper pool upstream of gates.
- Lower pool downstream of gates.
- Applicable wall loadings.

(b) Loading Condition 1B - Normal Operating Condition.

- Gates closed.
- For upper gate bay, upper pool in gate bay.
- For lower gate bay, lower pool in lock chamber.
- Applicable wall loadings.

(c) Loading Condition 2B - Operating Conditions with Extreme Low Water. The same requirements for conditions 1A and 1B except for the following conditions.

- Pools in lock chamber or lock entrance, with extreme low tailwater stages.
- Uplift as defined by water elevations.

(d) Loading Condition Case No=s. 2C and 2D - Operating Drains Ineffective Condition. The same requirements for Conditions 1A and 1B are included except for the condition of the raised saturation level caused by ineffective drains or ponding.

(e) Loading Condition 2E - Maintenance Condition. The same requirements as for Condition 1B except for the following conditions.

- Lock chamber unwatered to a predetermined level.
- Uplift as defined by water elevations.

(f) Loading Condition Case No=s. 2F and 2A- Normal Operating +OBE. The same requirements for Conditions 1A and 1B except for the condition of the OBE loads added in the most critical direction.

(g) Loading Condition Case No=s. 2G and 2H- Normal Operating + MDE. The same requirements for Cases 1A and 1B except for the condition of the MDE loads added in the most critical direction.

(h) Loading Condition 3A - Construction Conditions.

- Moist backfill to a predetermined level.
- Permanent or construction surcharge.
- Wind as applicable.
- No uplift.
- Gates swinging freely in appropriate mitered position.

- Hydrostatic forces are active in accordance with construction or cofferdam plans.

(3) Upper and lower approach walls.

(a) Loading Condition 1A - Usual Barge-Impact Loading.

- Water and backfill loads.
- Barge impact (Maximum mean impact force with a return period of 10 years) on face of wall at most critical angle of incidence.
- Uplift as defined by water elevations.

(b) Loading Condition 1B - Unusual Barge-Impact Loading. The same requirements as Condition 1A except the barge impact is a force representing the mean maximum force with a return period of 300 years.

(c) Loading Condition Case 1C - Extreme Barge-Impact Loading. The same requirements as Condition 1A except the barge impact is a force representing the mean maximum force with a return period of 1,000 years.

(d) Loading Condition 2A - Normal Operating + OBE.

- Most critical normal operating condition.
- OBE loads in the most critical direction.
- No impact or hawser pull.

(e) Loading Condition 2B - Normal Operating + MDE.

Most critical normal operating condition.

- MDE loads in the most critical direction.
- No impact or hawser pull.

(f) Loading Condition 3 - Construction Conditions.

- Moist backfill.
- Permanent or construction surcharge.
- Wind as applicable.
- No uplift.

(4) Upper and lower sills.

(a) Loading Condition 1 - Normal Operating Condition.

- Upper pool upstream of gate.
- Lower pool downstream of gate.
- Fill or silt to top of sill on upstream side.
- Applicable gate loads for vertically framed miter gates and rolling gates.
- Uplift and vertical water loading as defined by water elevations.

(b) Loading Condition 2A - Operating Condition with Extreme Low Tailwater. The same requirements for Case 1 are included except for the following conditions.

- Extreme low-water stage.

(c) Loading Condition 2B - Maintenance Condition.

- Upper pool upstream of temporary closure structure.
- Lock chamber unwatered.
- Uplift and vertical water loading as defined by water elevations.

Table B-11 Spillway Loading-Condition Classification
Structure Type: Spillway – Approach Channel Walls EM 1110-2-2400

Load Case	Loading Description	Classification
I	Channel Empty	U
II	Partial Sudden Drawdown, PMF	E
III	Sudden Pool Rise, PMF	E
IVA	Coincident Pool + OBE	UN
IVB	Coincident Pool + MDE	E

Table B-12 Spillway Loading-Condition Classification
Structure Type: Spillway – Chute Walls EM 1110-2-2400

Load Case	Loading Description	Classification
I	Channel Empty	U
II	Water in Channel, PMF	E
IIIA	Coincident Pool + OBE	UN
IIIB	Coincident Pool + MDE	E

Table B-13 Spillway Loading-Condition Classification
Structure Type: Spillway – Basin Walls EM 1110-2-2400

Load Case	Loading Description	Classification
I	Construction or Maintenance	UN
II	Rapid Closure of Gates	UN
III	MDF Discharge Conditions	U/UN/E
IVA	Coincident Pool + OBE	UN
IVB	Coincident Pool + MDE	E

Table B-14 Spillway Loading-Condition Classification
Structure Type: Spillway – Stilling Basin Floor EM 1110-2-2400

Load Case	Loading Description	Classification
I	MDF Discharge	U/UN/E
II	Rapid Closure of Gates	UN/E

Table B-15 Spillway Loading-Condition Classification
Structure Type: Spillway – Weir EM 1110-2-2400

Load Case	Loading Description	Classification
	Per Gravity Dam Requirements EM 1110-2-2200	

f. Spillways - basic loading conditions. Stability analyses for overflow sections of gravity and arch dams, and for spillways of embankment dams must not only investigate the maximum differential head conditions described for nonoverflow sections, (spillway gates closed), but also the maximum differential head and uplift conditions that can occur during spill (gates open). The spillway weir, chute walls, apron, and stilling basin all have different conditions that can cause maximum differential head, and maximum uplift. Conditions and combinations of loadings that will ordinarily require examination are described in the following paragraphs for each of the various types of structures common to spillways (Tables B11-15).

(1) Approach channel walls.

(a) Loading Condition I - Channel Empty, Pervious Drained Backfill Conditions.

- Channel empty.
- Backfill submerged to elevation of line of drains, and naturally drained above this elevation. Surcharge loading on backfill, if applicable.
- Uplift defined by water elevations.

(b) Loading Condition II - Partial Sudden Drawdown, Impervious Backfill Conditions.

- Partial sudden drawdown of reservoir from PMF elevation.
- Water in channel to drawdown elevation, which may occur suddenly.
- Fill submerged to profile reached during PMF, drained above.
- Uplift defined by water elevations.

(c) Loading Condition III - Sudden Rise of Reservoir, Impervious Backfill Conditions.

- Sudden rise of reservoir to PMF elevation.
- Water in channel to PMF conditions.
- Fill submerged to concurrent water surface in fill, naturally drained above.
- Water above fill to PMF elevation.
- Uplift defined by water elevations.

(d) Loading Condition IVA - Coincident Pool + OBE.

- Coincident pool elevation as defined in Paragraph B-2b(2).
- Backfill to predetermined height.
- Surcharge loading, if applicable.
- Uplift defined by water elevations.
- OBE loads in most critical direction.

(e) Loading Condition IVB - Coincident Pool + MDE. The same requirements as for Condition IVA except the MDE is used instead of the OBE.

(2) Spillway chute walls.

(a) Loading Condition I - Channel Empty, Pervious Drained Backfill Conditions.

- Channel empty.
- Backfill submerged to elevation of drains.
- Backfill naturally drained above elevation of drains.
- Surcharge loading on backfill, if applicable.
- Uplift defined by water elevations.

(b) Loading Condition II - Water in Channel to PMF Elevation.

- Water in channel to PMF conditions.
- Backfill submerged to elevation of drains.
- Backfill naturally drained above drains.
- Surcharge loading on backfill, if applicable.
- Uplift defined by water elevations.

(c) Loading Condition IIIA --Coincident Pool + OBE.

- Coincident pool elevation as defined in Paragraph B-2b(2).
- Backfill to predetermined height.
- Surcharge loading, if applicable.
- Uplift defined by water elevations.
- OBE loads in most critical direction.

(d) Loading Condition IIIB - Coincident Pool + MDE. The same requirements as for Case IIIA, except the MDE is used instead of the OBE.

(3) Stilling basin walls.

(a) Loading Condition 1- Construction or Maintenance Condition.

- Stilling basin empty.
- Backfill submerged to drain or higher if, during construction or maintenance, higher elevation is anticipated with stilling basin unwatered.
- Backfill above drain naturally drained.
- Surcharge, if applicable.
- Uplift defined by water elevations.

(b) Loading Condition II - Rapid Closure of Gates or Reduction of Discharge of Ungated Spillway.

- Maximum reduction of discharge and tailwater, which is expected to occur rapidly.
- Water surface inside stilling basin at tailwater corresponding to reduced flow conditions.
- Backfill submerged to elevation midway between tailwater before and after reduction (corresponding to 50-percent reduction by drainage).
- Backfill above level of submergence naturally drained.
- Uplift of uniform intensity across the base with pressure equal to reduced hydrostatic head in backfill.

(c) Loading Condition III - MDF Operating Conditions, Pervious Backfill.

- Water surface inside at hydraulic jump profile for MDF discharge condition. This condition creates the greatest differential head between outside and inside faces of the wall.
- Backfill submerged to the MDF tailwater conditions.

- Backfill above tailwater is naturally drained.
- Uplift across base varying uniformly from tailwater at heel to value midway between tailwater and jump profile at toe (the latter corresponds to 50-percent relief of unbalanced pressure by floor drainage).

(d) Loading Condition IVA - Coincident Pool + OBE.

- Coincident pool elevations as defined by Paragraph B-2b(2).
- Backfill to predetermined height.
- Surcharge loading, if applicable.
- Uplift defined by water elevations.
- OBE loads in most critical direction.

(e) Loading Condition IVB - Coincident Pool + MDE. The same requirements as for Condition IVA except the MDE is used instead of the OBE.

(4) Stilling basin floor.

(a) Loading Condition I - Stilling Basin Operating During MDF.

- MDF condition creating greatest differential pressure between bottom and top surfaces of the basin floor.
- Water surface over slab at hydraulic jump profile.
- Hydrostatic pressure under slab to elevation of full tailwater depth.

(b) Loading Condition II - Rapid Closure of Gates or Reduction of Discharge of Ungated Spillway.

- Maximum probable reduction of discharge and tailwater inside stilling basin due to gate closure.
- Water surface inside stilling basin at tailwater corresponding to reduced flow conditions.
- Uplift of uniform intensity with pressure equal to hydrostatic head midway between tailwater before and after reduction (corresponding to 50-percent reduction by drainage).

Table B-16 Power Plant Loading-Condition Classification
Structure Type: Power Plant Structures EM 1110-2-3001

Load Case	Loading Description	Classification
S or M-1A	Normal Operating Condition	U
S or M-1B	Flood-Control Pool Condition	UN/E
S or M-2	Powerhouse Flooding Condition	UN/E
S or M-3	Draft Tube Flood Condition	UN/E
S or M-4	Construction Condition	UN
S or M-5A	Coincident Pool + OBE	UN
S or M-5B	Coincident Pool + MDE	E

g. Power plant structures - basic loading conditions. Powerhouses may either be part of a dam or may be located separate from the dam. In either condition, stability analyses must consider the maximum differential head and uplift conditions that occur with the gates closed and the water passages empty, as well as those that can occur with the gates open and the water passages full. A stability analysis should be made for each monolith of the powerhouse, and all critical levels should be investigated for the most severe combinations of horizontal and vertical forces. In the condition of a monolith in which the power unit will not be installed with the initial construction, the stability analysis should be investigated for the interim as well as the final condition. Analysis should be made for the applicable conditions indicated below and for any other combinations of conditions, which might prove critical.

Conditions S-1A, S-1B, S-2, S-3, S-4, S-5A, and S-5B are applicable when the powerhouse is separated from the dam, and Conditions M-1A, M-1B, M-2, M-3, M-4, M-5A, and M-5B are applicable when the powerhouse and headworks form a part of the dam (Table B-16).

(1) Powerhouse separated from the dam.

(a) Loading Condition S-1A - Normal Operating Condition.

- Head gates open, headwater at normal operating pool level as defined by Paragraph B-2b(1).
- Hydraulic thrusts.
- Minimum tailwater.
- Spiral case full.
- Draft tube full.
- Uplift.

(b) Loading Condition S-1B - Flood-Control Pool Operating Condition.

- Head gates open, headwater at flood-control pool elevation.
- Hydraulic thrusts.
- Minimum tailwater.
- Spiral case full.
- Draft tube full.
- Uplift .

(c) Loading Condition S-2 - Powerhouse Flooding Condition.

- Head gates open.
- Tailwater at powerhouse flooding level.
- Spiral case full.
- Draft tube full.
- Uplift.

(d) Loading Condition S-3 - Draft Tube Flooding Condition.

- Head gates closed.
- Tailwater at draft-tube flooding level.
- Spiral case empty.
- Draft tube empty.
- Uplift.

(e) Loading Condition S-4 - Construction Condition.

- No tailwater.
- No uplift .

(f) Loading Condition S-5A - Coincident Pool Condition + OBE.

- Head gates open, headwater at coincident pool elevation as defined by Paragraph B-2b(2).
- Hydraulic thrusts.
- Minimum tailwater.
- Spiral case full.
- Draft tube full.
- Uplift.
- OBE loads acting in most critical direction.

(g) Loading Condition S-5B - Coincident Pool Condition + MDE.

- The same requirements for S-5A except for the condition of an MDE, rather than OBE, acting in the most critical direction.

(2) Powerhouse part of the dam.

(a) Loading Condition M-1A - Normal Operating Condition.

- Head gates closed, headwater at normal operating pool level as defined by Paragraph B-2b(1).
- Minimum tailwater.
- Spiral case open to tailwater.
- Draft tube open to tailwater.
- Uplift.

(b) Loading Condition M-1B - Flood-Control Pool Operating Condition.

- Head gates open, headwater at flood-control pool elevation.
- Minimum tailwater.
- Spiral case open to tailwater.
- Draft tube open to tailwater.
- Uplift .

(c) Loading Condition M-2 - Powerhouse Flooding Condition.

- Head gates open.
- Headwater at maximum flood level.
- Tailwater at powerhouse flooding level.
- Spiral case full.
- Draft tube full.
- Uplift.

(d) Loading Condition M-3 - Draft Tube Flooding Condition.

- Head gates closed.
- Headwater at top of flood-control pool.
- Tailwater at draft-tube flooding level.
- Spiral case empty.
- Draft tube empty.
- Uplift.

(e) Loading Condition M-4 - Construction Condition.

- No uplift.

(f) Loading Condition M-5A - Coincident Pool Condition + OBE.

- Head gates closed, headwater at coincident pool elevation as defined by Paragraph B-2b(2).
- Minimum tailwater.
- Spiral case open to tailwater.
- Draft tube open to tailwater.
- Uplift.
- OBE loads acting in most critical direction.

(g) Loading Condition M-5B - Coincident Pool Condition + MDE. The same requirements for M-5A except for the condition of an MDE, rather than OBE, acting in the most critical direction.

Table B-17 Pumping Station Loading-Condition Classification
Structure Type: Pumping Stations, EM 1110-2-3104

Load Case	Loading Description	Classification
4-4a	Construction Condition	UN
4-4b	Normal Operating Condition	U
4-4e	MDF	U/UN/E
4-4f	Maximum Pump Thrust	U/UN/E
4-4g	Maintenance Condition	UN
4-4j	Pumping Station Inundated	E
4-4k	Coincident Pool + OBE	UN
4-4l	Coincident Pool + MDE	E

h. Pumping stations - basic loading conditions. The loading conditions listed in Table B-17 and described below should not be regarded as a comprehensive list. In many instances, unique, site- specific factors such as water conditions, station arrangement and location, pump type and discharge arrangement, etc. will dictate modification of some of these loading conditions to fit the specific site. The conditions described should be used as a guide to the range of stability analyses required.

(1) Design loadings for stability - pumping stations.

(a) Loading Condition 4-4a - Construction Condition.

- Pumping station complete with and without backfill in place.
- No water loads.

(b) Loading Condition 4-4b - Normal Operating Condition.

- Plant operating to discharge routine local floods over a range of exterior flood levels with a maximum 2-year return period.

(c) Loading Condition 4-4e - MDF.

- Maximum water level outside protection line.
- Minimum pumping level inside.

(d) Loading Condition 4-4f - Maximum Pump Thrust.

- Maximum operating floods both inside and outside protection line.
- Maximum pump thrust.

(e) Loading Condition 4-4g - Maintenance Conditions.

- Maximum design water level inside.
- One, more, or all intake bays unwatered.

(f) Loading Condition 4-4j - Pumping Station Inundated.

- Maximum flood levels inside and outside protection line.
- Pumping station inoperative.
- Foundation drains inoperative.
- Protection line intact.

(g) Loading Condition 4-4k - Coincident Pool + OBE.

- Coincident pool
- OBE in most critical direction.

(h) Loading Condition 4-4l- Coincident Pool + MDE.

- Coincident pool
- MDE in most critical direction.

Table B-18 Concrete-Lined Flood-Control Channels
Structure Type: Retaining Wall and U-frame Structures EM 1110-2-2007

Load Case	Loading Description	Classification
1	Construction Condition	UN
2	Design Flood Loading	U/UN/E
3	Drawdown Loading	U
4a	Normal Operating + OBE	UN
4b	Normal Operating + MDE	E

i. Concrete-lined flood-control channels - basic loading conditions.

(1) Loading Condition 1 - Construction Condition (Table B-18)

- Structure complete and backfill in place.
- Channel empty.
- Construction surcharge loadings.

(2) Loading Condition 2 - Design flood loading.

- Structure complete and backfill in place.
- Water level at maximum design flood level.
- Backfill saturated to low groundwater level, adjusted to reflect the design effectiveness of the drainage system.

(3) Loading Condition 3 - Drawdown Loading.

- Structure complete and backfill in place.
- Channel empty.
- Hydrostatic pressures reflecting the highest groundwater level, adjusted to reflect the design effectiveness of the drainage system.

(4) Loading Condition 4a - Normal Operating + OBE

- Structure and backfill in place.
- Channel at mean annual operating water level.
- Backfill saturated to normal groundwater level, adjusted to reflect the design effectiveness of the drainage system.
- OBE loads acting in most critical direction.

(5) Loading Condition 4b - Normal Operating + MDE.

- The same requirements for Condition 4a except for the condition of an MDE, rather than OBE, acting in the most critical direction.

Appendix C
Uplift

C-1. General

This appendix provides a summary of design uplift requirements as they apply to gravity dams, spillway chutes, navigation locks, and other miscellaneous structures. These design uplift requirements will produce conservative designs under most geologic site conditions. The permeability of the foundation soils, or for rock foundations, the permeability of joints, faults, and shear zones in the foundation, greatly affect uplift pressures. Therefore, close coordination with geotechnical engineers is needed in determining uplift pressures. Drainage can be used effectively to reduce uplift pressures. The uplift pressure at any point under the structure will be dependent on the presence, location, and effectiveness of foundation drains. Cutoffs such as grout curtains, impervious blankets, sheet-pile walls, and keys also affect uplift pressures and should be considered in determining design uplift pressures and drainage requirements. Seepage flow net and creep theory can be used to determine uplift pressures for structures on soil foundations. The fundamental design principles and guidance concerning seepage are detailed in EM 1110-2-1901. Uplift pressure is an applied force that must be included in the stability and stress analysis. The uplift pressure will be considered as acting over 100 percent of the base. Uplift pressures are assumed to be unchanged by earthquake loads. Uplift assumptions are valid only if there is adequate resistance to piping. If there is a concern about piping, geotechnical engineers should be consulted.

C-2. Uplift Pressures Due To Seepage

a. *General.* Where seepage occurs, the pressure heads at points of interest must be obtained from a seepage analysis. Where soil conditions adjacent to and below a structure can be assumed homogeneous (or can be mathematically transformed into equivalent homogeneous conditions), simplified methods such as the line-of-seepage method may be used. However, designers should ensure that water pressures are based on appropriate consideration of actual soil conditions. The line-of-seepage method is illustrated in Figure C-1. The uplift pressures at the ends of the base (points B and C) are estimated by assuming that the head varies linearly along the shortest possible seepage path (ABCD). Where a key is present (Figure C-2), point B is at the bottom of the key, and line BC is drawn diagonally. Permeabilities that are different in the horizontal and vertical directions can be handled by adjusting the length of the different segments along the total seepage path in accordance with the relationship between these different permeabilities.

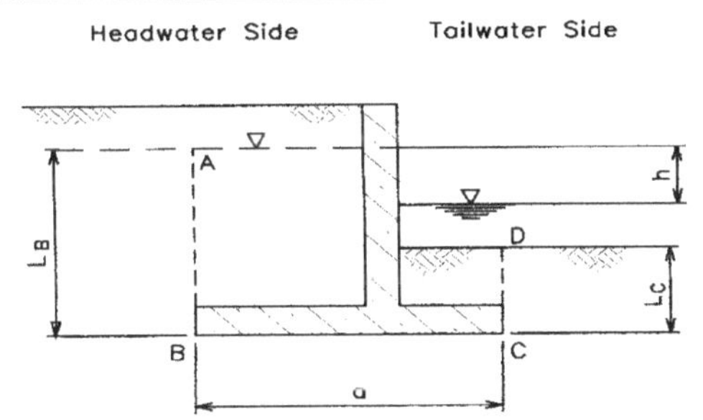

L = Total length of seepage path = $L_B + a + L_C$

h = Head difference across wall

u_B = Water pressure at B $= \left[L_B - \frac{h\,L_B}{L}\right]\gamma_w$

u_C = Water pressure at C $= \left[L_B - \frac{h(L_B + a)}{L}\right]\gamma_w$

Where γ_w = The unit weight of water

Figure C-1 Line-of-seepage method for water pressures

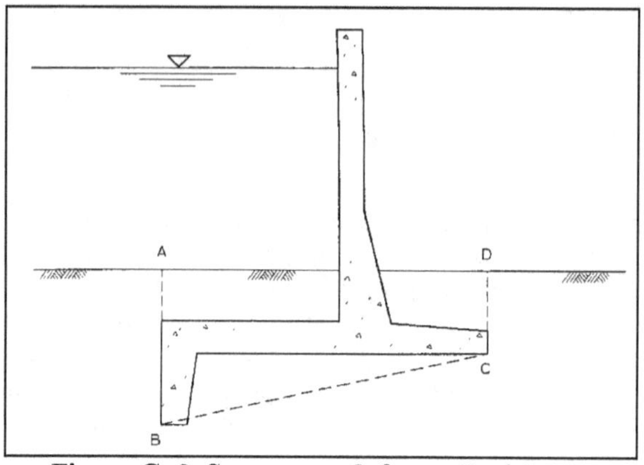

Figure C- 2 Seepage path for wall with key

b. *Uplift calculation for rock foundations.* Seepage beneath structures founded on rock typically occurs in joints and fractures, not uniformly through pores as assumed for soils. Consequently, the assumptions of isotropy and homogeneity and the use of two-dimensional analysis models employed for soil foundations will generally be invalid. Total head, uplift pressure, and seepage quantities may be highly dependent on the type, size, orientation, and continuity of joints and fractures in the rock and the type and degree of treatment afforded the rock foundation during construction. For structures on rock, the total seepage path can be assumed to be length of base that is in compression.

c. *Effect of gallery drains on uplift pressures.* Drainage is an effective means of reducing uplift pressures for structures founded on rock. It also is effective for structures founded on soils, provided the loss of soil materials through piping can be prevented. For a dam with a line of drains connected to a drainage gallery, uplift downstream of the line of drains can be at or near tailwater, provided the drainage gallery floor is at or below tailwater, the drains are adequately spaced, and they penetrate the pervious strata. For design, it would be unconservative to assume that the aforementioned ideal drainage conditions exist. Normally for design purposes, the drain efficiency is assumed to be 50 percent. Since drainage is such an important factor in reducing uplift pressures, the best policy is to regularly inspect, maintain, and clean the drains to prevent clogging. In cases where it is impossible to clean the drains, a drains-clogged condition should be included as part of the stability analysis for an existing structure. This load condition should be treated as a usual, unusual, or extreme load condition depending on the return period for the load condition being evaluated.

d. *Effect of cutoffs on uplift pressures.* Cutoffs can also contribute to reducing uplift below structures. Cutoffs can be either grout curtains, concrete trenches, steel sheet piling, or impervious blankets. The effectiveness of cutoffs, however, can be jeopardized by leakage through joints, cracks, and fractures. Therefore, drains are considered to be the most reliable and cost-effective way of reducing foundation-uplift pressures, especially for structures founded on rock. Although grout curtain cutoffs are commonly used in combination with drainage systems for dams founded on rock, the grout-curtain cutoff helps more to reduce drain flows in the drainage gallery than to reduce uplift pressures. Steel sheet-pile cutoffs are not entirely watertight due to leakage at the interlocks, but can significantly reduce the possibility of piping of coarse-grained foundation material. The efficiency of a steel sheet- pile cutoff through a coarse-grained stratum in reducing uplift depends upon conditions at the interlocks, the penetration distance (P) of the cutoff into the pervious stratum, and the depth (D) below the base of the structure to the top of impervious material. When $P \geq 0.95D$, and the pile interlocks are in good condition, an efficiency (E) of 0.50 may be assumed. It has been observed that some steel sheet piling driven into certain types of foundation material (such as gravel) can experience a complete loss of interlock. Before relying on a steel sheet-pile cutoff the designer must be certain that the assumed reduction in seepage will actually occur. A design uplift diagram and method for computing pressures at pertinent points are shown in Figure C-3. In this figure, uplift pressure for that part of the base on the heel side of the cutoff is that due to the full head (no seepage). Seepage is assumed to occur between the cutoff and the toe of the structure, and soil permeabilities are assumed equal in the horizontal and vertical directions.

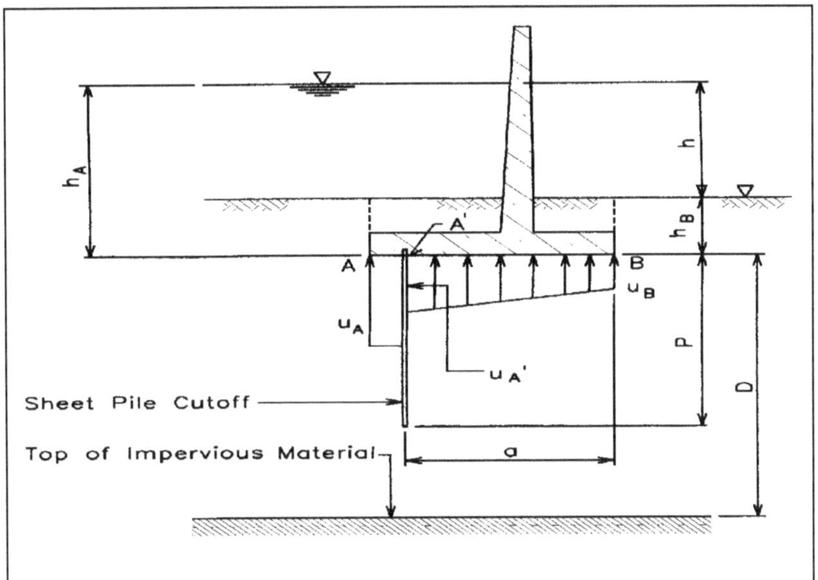

u_A = uplift at A = $\gamma_w h_A$

u_B = uplift at B = $\gamma_w h_B + \frac{h_B}{a+h_B}(u_{A'} - \gamma_w h_B)$

$u_{A'}$ = uplift at A' = $u_A - E(u_A - \gamma_w h_B)$

Where:

γ_w = The unit weight of water

E = Efficiency of sheet pile cutoff

Figure C-3 Uplift pressures with sheet pile cutoff

C-3. Design Uplift for Gravity Dams

a. General. A hydraulic gradient between the upper and lower pool is developed between the heel and toe of the dam. The pressure distribution along the base and in the foundation is dependent on the effectiveness of drains and cutoffs, where applicable, and geologic features such as rock permeability, seams, jointing, and faulting.

b. Without drains. Where there are no provisions for uplift reduction, the hydraulic gradient will be assumed to vary, as a straight line, from headwater at the heel to zero or tailwater at the toe, assuming the entire base remains in compression between the concrete and the foundation.

c. With drains. Uplift pressure at the base, or below the foundation, can be reduced by installing foundation drains. The effectiveness of the drainage system will depend on the depth, size, and spacing of the drains; the character of the foundation; and the facility with which the drains can be maintained. The assumed effectiveness will be limited to no greater than 50 percent, and the design documentation should contain supporting data to justify this assumption. If foundation testing and flow analysis provide supporting justification, the drain effectiveness can be increased beyond 50 percent. Use of a higher assumed drain effectiveness will depend on the pool-level operation plan, instrumentation to verify and evaluate uplift assumptions, and an adequate drain maintenance program. Along the base, the uplift pressure will be assumed to vary linearly from the undrained pressure head at the heel, to the reduced pressure head at the line of drains, to the undrained pressure head at the toe, as shown in Figure C-4. This figure also provides formulas for uplift calculation where the drainage gallery is above or below tailwater.

d. Grout curtain. For drainage to be controlled economically, retarding flow to the drains from the upstream head is mandatory. This may be accomplished by a zone of grouting (curtain) or by the natural imperviousness of

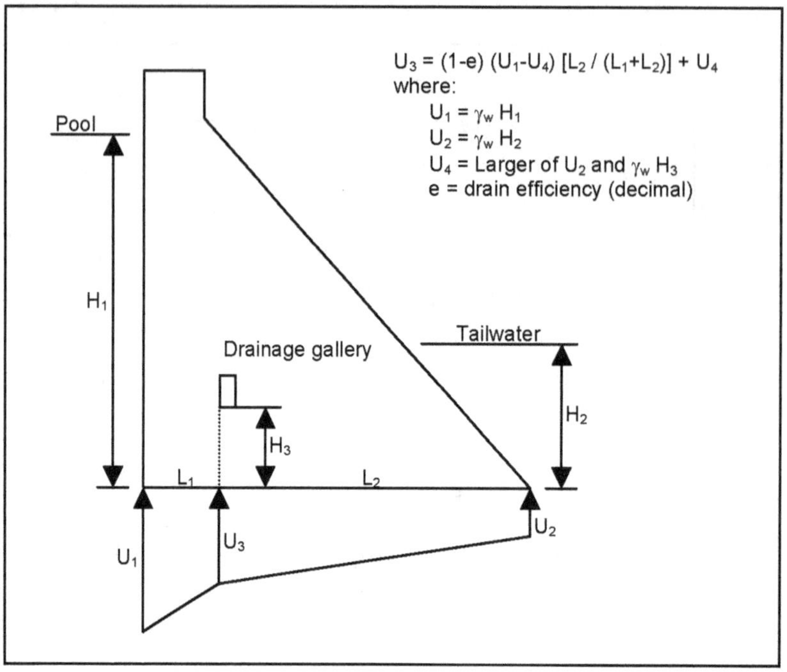

Figure C-4 Uplift distribution with foundation drains

the foundation. A grout curtain should be used wherever the foundation is amenable to grouting. Grout holes should be oriented to intercept the maximum number of rock fractures to maximize its effectiveness. Under average conditions, the depth of the grout zone should be two-thirds to three-fourths of the headwater-tailwater differential and should be supplemented by foundation drain holes with a depth of at least two-thirds that of the grout zone. Where the foundation is sufficiently impervious to retard the flow and where grouting would be impractical, an artificial cutoff is usually unnecessary. Drains, however, should be provided to relieve the uplift pressures that would build up over a period of time in a relatively impervious medium. In a relatively impervious foundation, drain spacing will be closer than in a relatively permeable foundation.

e. Zero compression zones. Any portion of the foundation plane not in compression shall be assumed to behave as an open crack, except where loss of compression is the result of instantaneous loading resulting from earthquake forces. For the length of this crack, uplift pressures shall be assumed as 100 percent of the hydrostatic head of the adjacent face. When the zero compression zone does not extend beyond the location of the drains, the uplift will be calculated using the formula in Figure C-4, but substituting the length of the compression zone as if it was the full length of the base. When the zero compression zone extends beyond the drains, drain effectiveness shall be assumed to be negligible. Uplift shall vary linearly from headwater at the end of the crack, to tailwater at the toe of the dam. When an existing dam is being evaluated, and expensive remedial measures are required to satisfy this loading assumption, the design office should consult with CECW-E to determine if a waiver would be appropriate.

f. Overflow sections. For overflow sections, tailwater pressure must be adjusted for retrogression when flow conditions result in a significant hydraulic jump in the downstream channel, i.e., spillway flow plunging deep into tailwater. The forces acting on the downstream face of overflow sections due to tailwater may fluctuate significantly as energy is dissipated in the stilling basin. Therefore, these forces must be conservatively estimated when used as a stabilizing force in a stability analysis. Studies have shown that the influence of tailwater retrogression can reduce the effective tailwater depth used to calculate pressures and forces to as little as 60 percent of the full tailwater depth. The amount of reduction in the effective depth used to determine tailwater forces is a function of the degree of submergence of the crest of the structure and the backwater conditions in the downstream channel. For new designs, EM 1110-2-1603 provides guidance in the calculation of hydraulic pressure distributions in spillway flip buckets due to tailwater conditions. When tailwater conditions significantly reduce or eliminate the hydraulic jump in the stilling basin, tailwater retrogression can be neglected and 100 percent of the tailwater depth can be used to determine tailwater forces. Full tailwater depth will be used to calculate uplift pressures at the toe of the structure in all cases, regardless of overflow conditions. Figure C-5 illustrates the forces and uplift pressures to be used in stability analysis for an overflow and stilling basin section operating under hydraulic jump conditions.

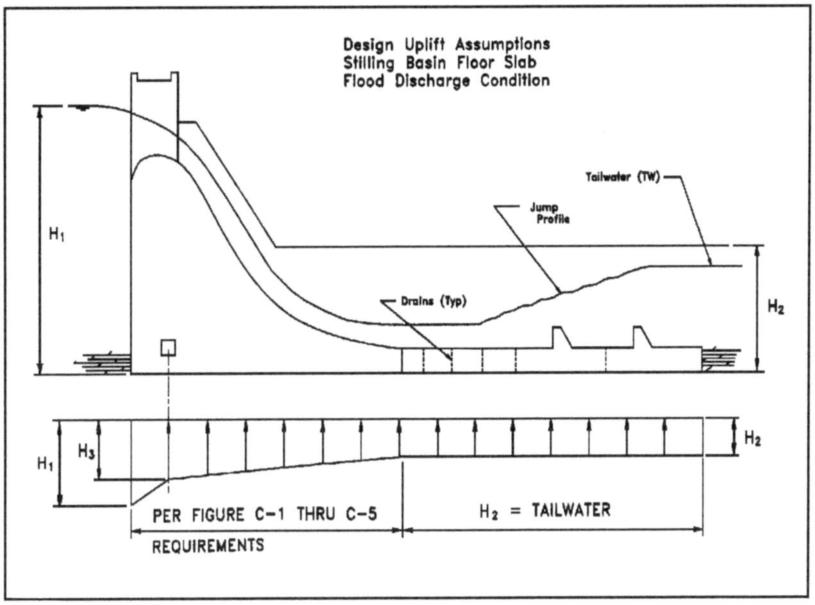

Figure C-5 Stilling basin with discharge condition

C-4. Design Uplift for Navigation Locks

a. General. The problem of uplift for lock walls is complicated by fluctuating water levels within a lock chamber. The rate of change of uplift as the chamber is filled or emptied is not known. The design uplift assumptions used for the stability analysis of lock structures is similar to gravity dams and is described in the following paragraphs.

b. Rock foundations. During construction, uplift acting on the base of any monolith within the cofferdam is assumed to be zero. For walls without drains, uplift will vary linearly from the chamber face to the opposite face; uplift at each face will be 100 percent of the water elevation adjacent to that face (either the current chamber water elevation, the river water elevation or the saturation line in the backfill). In cases where adequate drainage (relieving to tailwater) is provided near the chamber face, total uplift may be reduced for the condition of upper pool in the lock chamber. For river walls, uplift will vary from 100 percent of tailwater plus 50 percent of the difference between headwater and tailwater at the chamber face to 100 percent of tailwater at the river face. For land walls, use the saturation line instead of tailwater. Probably the most effective land wall drainage is that provided in the backfill to reduce the saturation level.

c. Soil and pile foundations. Monoliths on soil or pile foundations usually have cutoff walls and sometimes have drainage systems. At one face of the monolith, uplift should be the full headwater pressure from the face of the wall to the cutoff. At the other face, uplift equals the full tailwater pressure (or the saturation head in the backfill). Uplift pressures between these points should be determined by evaluations of cutoff and drain effectiveness and soil permeability. Cutoffs and drains will normally be designed for 50 percent reduction in uplift, similar to rock foundations. Under excellent conditions, cutoffs and drains can be considered beyond 50 percent effective in reducing uplift pressures, subject to approval from CECW-E. Except for earthquake loading, any portion of the base not in compression will be assumed to sustain a uniform uplift equivalent to 100 percent of the adjacent pool or saturation level. Uplift for loadings, which include earthquake forces, will be assumed to be equal to that for the same loading without earthquake forces. Because minor movements of gate sills affect the gate operations, all sill blocks should be analyzed for stability resulting from maximum differential heads. Uplift on the sills should be determined similar to the lock walls. The uplift under U-frame locks is complicated by alternative seepage paths along and perpendicular to the lock axis. The permeability of the foundation soils, as well as the existence of sheet-pile cutoff walls and foundation drains, affect the magnitude and distribution of the uplift pressure. Close coordination with geotechnical engineers is needed to determine the uplift pressure for each lock monolith. All combinations of operating and maintenance conditions should be analyzed to determine the most critical condition.

C-5. Design Uplift for Other Structures

The influence that drains and cutoffs have on uplift pressures for other structures such as retaining walls, intake towers, and lined flood-control channels are similar to those described above for gravity dams and navigation locks. A drainage system should be considered for all retaining walls. The benefits of a drainage system are a reduction in hydrostatic pressures in the backfill, lower lateral water pressures on the structure, and a reduction of uplift pressures.

Appendix D
Example Problems

The examples provided in this appendix clearly demonstrate the procedure for applying the multiple-wedge and single-wedge equations to the stability analysis of wedge systems, stability benefits from uplift reduction, stability benefits of tensioned structural anchors, obtaining shear strength parameters for use in sliding-stability analysis, uplift pressures and differential heads on stilling-basin slabs, and calculating critical wedge-slip angles for selected conditions. The variations of uplift pressure, orientation of failure planes, etc., used in the examples were only selected to simplify the calculations and are not intended to represent the only conditions to be considered during design or investigation of the stability of a hydraulic structure. Dimensions, weights, volumes, densities, forces, and pressures are presented in SI units, with corresponding American units given inside brackets.

D-1. Example D1, Gravity-Dam Sliding Analysis, Single-Wedge System

Determine the factor of safety against sliding for the structure shown in Figure D-1, using the single-wedge sliding analysis method as discussed in Chapter 5 of this EM. As a point of interest, the multiple-wedge analysis would produce the same results for this problem.

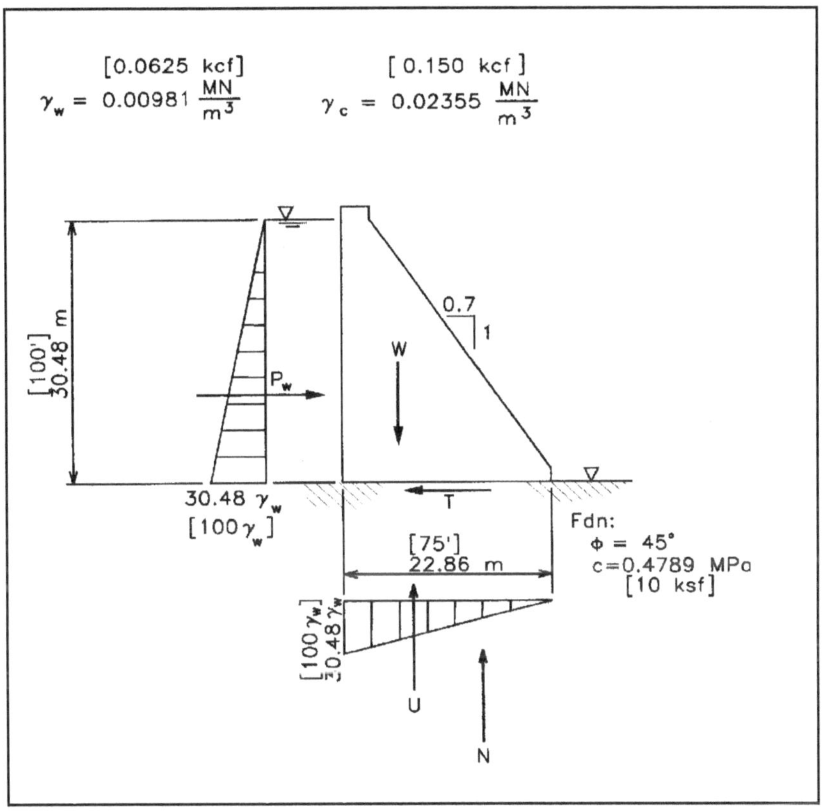

Figure D-1

$$W = 8.812 MN/m [603.8 k/ft]$$

$$T = P_w = \frac{1}{2}\gamma_w h^2 = \frac{1}{2}(0.00981)(30.48)^2 = 4.557\frac{MN}{m}, [312.5\text{ k}]$$

$$U = \frac{1}{2}\gamma_w hb = \frac{1}{2}(0.00981)(30.48)(22.86) = 3.418\frac{MN}{m}\ [234.38\ \text{k/ft}]$$

$$N = W - U = 8.812 - 3.418 = 5.394\frac{MN}{m}\ [369.42\ \text{k/ft}]$$

$$FS = \frac{N\tan\phi + c\,L}{T} = \frac{5.394 \times 1 + 0.4789 \times 22.86}{4.557} = 3.59$$

D-2. Example D2,Gravity-Dam Sliding Analysis, Multiple-Wedge System

Determine the factor of safety against sliding for the gravity dam shown in Figure D-2. Use the multiple-wedge-sliding analysis as discussed in Chapter 2. The purpose of this example is to illustrate the calculations required for a multiple-wedge sliding analysis using the iterative method.

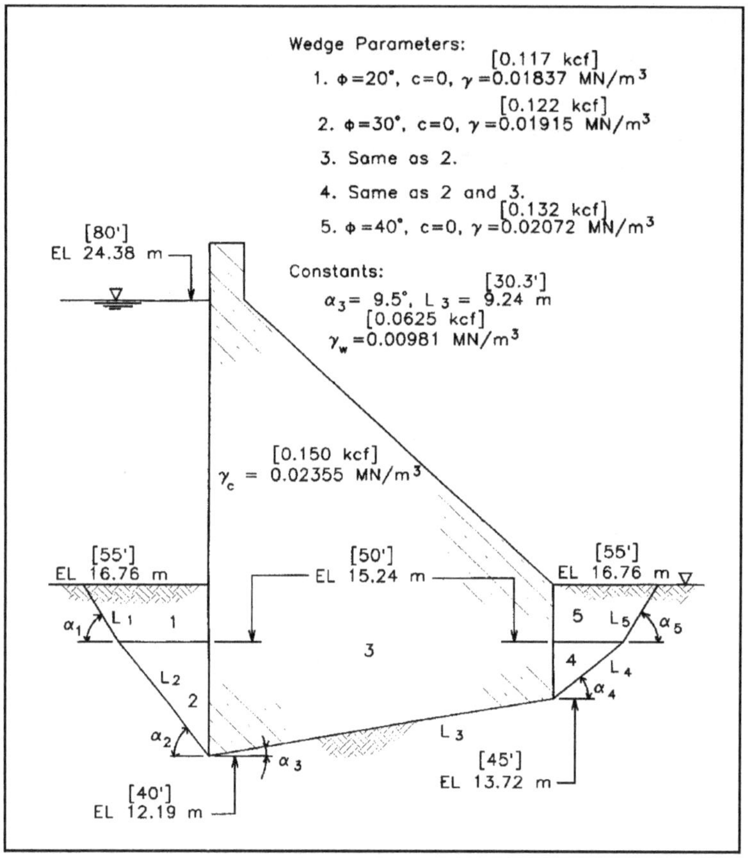

Figure D-2

For an usual load condition, a critical structure, and ordinary site information the minimum-required factor of safety is 2.0. (See Chapter 3).

$$P_{i-1}-P_i=\frac{[(W_i+V_i)\cos\alpha_i-U_i+(H_{Li}-H_{Ri})\sin\alpha_i]\frac{\tan\phi_i}{FS_i}-(H_{Li}-H_{Ri})\cos\alpha_i+(W_i+V_i)\sin\alpha_i+\frac{c_i}{FS_i}L_i}{\cos\alpha_i-\frac{\tan\phi_i\sin\alpha_i}{FS_i}}$$

Wedge forces for trial safety factor of 1.5

$\underline{i=1}\; H_{Li}=H_{Ri}=0,\quad V_1=0$

$$\tan\phi_d=\frac{\tan\phi_1}{FS_1}=\frac{\tan 20}{1.5}\qquad \phi_d=\tan^{-1}(0.243)=13.64°,\quad c_d=0$$

$$\alpha_1=-(45°+\frac{\phi_d}{2})=-51.82°$$

(This orientation is only true if the stratification and surface are horizontal.)

$\sin(-51.82)=-0.786,\quad \cos(-51.82)=0.618$

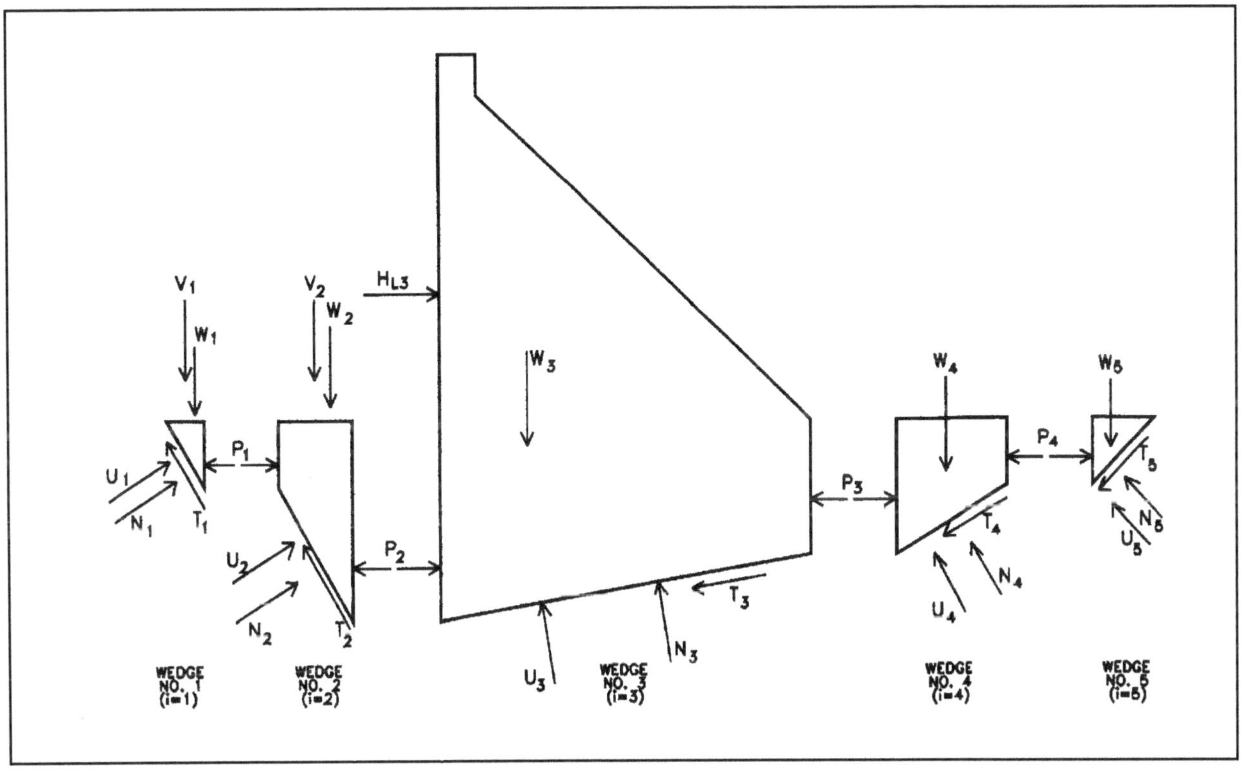

Figure D-3

$$L_1=1.52\,/\,[\,\sin(-51.82)]=1.52\,/\,0.786=1.934\ m\ [6.3'6\ ]$$

$$W_1 = \frac{1}{2}(0.00856)(1.52)(1.934)\cos(-51.82) = 0.0078\ MN/m\ [0.54\ \text{k/ft}]$$

$$(P_0 - P_1) = \frac{[(0.0078+0)(0.618) - (0-0)(-0.786)]\frac{\tan 20}{1.5} - (0-0)(0.618) + (0.0078+0)(-0.786) + \frac{0(1.934)}{1.5}}{0.618 - (-0.786)\frac{\tan 20}{1.5}} = -0.006\ MN/m\ [-0.42\text{k/}$$

$\underline{i=2}$ $H_{L2} = H_{R2} = 0$

$$\tan\phi_d = \frac{\tan\phi_2}{1.5} = \frac{\tan 30}{1.5} \qquad \phi_d = \tan^{-1}(0.385) = 21.05\ °$$

$$\alpha_2 = -(45 + \frac{\phi_d}{2}) = -55.53°$$

$$L_2 = \frac{3.05}{0.8244} = 3.70\ m\ [12.13\ \text{ft}]$$

$$W_2 = 0.00856(1.52)(3.70x0.566) + \frac{1}{2}(0.00934)(3.05)(3.70x0.566) = 0.0571MN/m\ [3.91\text{k/ft}],\ \ V_2 = 0,\ \ c_2 = 0$$

$$\sin(-55.53) = -0.8244 \quad \cos(-55.53) = 0.566$$

$$(P_1 - P_2) = \frac{[(0.0571+0)(0.566) - 0 + 0(-0.8244)]\frac{\tan 30}{1.5} - 0(0.566) + (0.0571+0)(-0.8244) + \frac{0(3.70)}{1.5}}{0.566 - (-0.8244)\frac{\tan 30}{1.5}}$$

$$(P_1 - P_2) = -0.040\ MN/m\ [-2.69\ \text{k/ft}]$$

$\underline{I=3}$ $\alpha_3 = 9.5°$ $L_3 = 9.270\ m\ [30.3\ \text{ft}]$, $c_3 = 0$, $V_3 = 0$, $U_3 = 0.6925$ MN/m [47.34 k/ft]

$$H_{L3} = \frac{1}{2}(0.00981)(12.19)^2 = 0.7289MN/m\ [49.95\ \text{k/ft}]$$

$$H_{R3} = \frac{1}{2}(0.00981)(3.04)^2 = 0.0453MN/m\ \ [3.11\ \text{k/ft}]$$

$$H_{L3} - H_{R3} = 0.6836\ \ MN/m\ \ [46.85\ \ \text{k/ft}]$$

$W_3 = 1.7864\ MN/m\ [122.4$ k/ft], $\sin 9.5° = 0.165$, $\cos 9.5° = 0.986$

$$(P_2 - P_3) = \frac{[(1.7864+0)(0.986)+0.6836(0.165)-0.6925]\dfrac{\tan 30}{1.5}}{0.986-0.165\ x\dfrac{\tan 30}{1.5}}$$

$$\frac{-0.6836(0.986)+(1.7864+0)(0.165)+\dfrac{0(9.27)}{1.5}}{0.986-0.165\,x\dfrac{\tan 30}{1.5}} = 0.082\,MN/m\ [5.61\ \text{k/ft}]$$

$\underline{i=4}\ H_{L4} = H_{R4} = V_4 = 0,\quad c_4 = 0$

$$\tan\phi_d = \frac{\tan\phi_4}{FS_4} = \frac{\tan 30°}{1.5}\qquad \phi_d = \tan^{-1}(0.385) = 21.05°$$

$$\alpha_4 = 45 - \frac{\phi_d}{2} = 34.475°\ \ \sin(34.475) = 0.566\ \ \cos(34.475) = 0.824$$

$$L_4 = \frac{1.52}{\sin 34.475} = 2.685\ m\ [8.8'3]$$

$$W_4 = (0.01091)(1.52)(2.685\ x\ 0.824) + \frac{1}{2}(0.00934)(1.52)(2.685\ x\ 0.824) = 0.0524\ MN/m\ [3.59\ \text{k/f}$$

$$(P_3 - P_4) = \frac{[(0.0524-0)(0.824)-0+0(0.566)]\dfrac{\tan 30}{1.5}-0(0.824)+(0.0524+0)(0.566)+\dfrac{0(2.865)}{1.5}}{0.824-0.566\left(\dfrac{\tan 30}{1.5}\right)}$$

$(P_3 - P_4) = 0.076\ MN/m\ [5.23\ \text{k/ft}]$

$\underline{i=5}\ H_L 5 = H_{R5} = V_5 = 0,\quad c_5 = 0$

$$\tan\phi_d = \frac{\tan\phi_5}{FS_5} = \frac{\tan 40}{1.5}\qquad \phi_d = \tan^{-1}(0.559) = 29.22°$$

$$\alpha_5 = 45 - \frac{\phi_d}{2} = 30.38°\quad \sin 30.38 = 0.5058\quad \cos 30.38 = 0.8626$$

$$L_5 = \frac{1.52}{\sin 30.38} = 3.01\,m\ [9.89\ \text{ft}]$$

$$W_5 = \frac{1}{2}(0.01091)(1.52)(3.01 - 0.8626) = 0.0215\ MN/m\ [1.48\ \text{k/ft}]$$

$$(P_4 - P_5) = \frac{[(0.0215+0)0.862)-0+0(0.506)]\frac{\tan 40}{1.5} - 0(0.863)+(0.0215+0)(0.506)+\frac{0(3.01)}{1.5}}{0.863-0.506\left(\frac{\tan 40}{1.5}\right)}$$

$$(P_4 - P_5) = 0.036\ MN/m\ [2.51\ \text{k/ft}]$$

Summing the resultant horizontal forces for all wedges:

$$(P_0-P_1)+(P_1-P_2)+(P_2-P_3)+(P_3-P_4)+(P_4-P_5) = -0.006-0.040+0.082+0.076+0.036 = 0.148 > 0$$

Since this summation is greater than zero, the safety factor (FS) must be greater than 1.5. Calculations have been made for trial safety factors of 2.5 and 2.0 in a similar manner, and the results are presented in Tables D-2 through D-4 and on a graph.

Table D-2 FS = 1.5

i	α	L	H_L	H_R	V	W	U	$P_{I-1} - P_i$
1	-51.82	1.934	0	0	0	0.0078	0	-0.006
2	-55.53	3.700	0	0	0	0.0571	0	-0.040
3	9.5	9.270	0.7289	0.0453	0	1.7864	0.6925	0.082
4	34.47	2.685	0	0	0	0.0524	0	0.076
5	30.38	3.010	0	0	0	0.0215	0	0.036

$\sum \Delta P = 0.148$

Table D-3 FS = 2.5

I	α	L	H_L	H_R	V	W	U	$P_{i-1} - P_i$
1	-49.14	2.015	0	0	0	0.0086	0	-0.007
2	-51.50	3.895	0	0	0	0.0661	0	-0.053
3	9.5	9.270	0.7289	0.0453	0	1.7864	0.6925	-0.112
4	38.50	2.438	0	0	0	0.0452	0	0.057
5	36.72	2.609	0	0	0	0.0169	0	0.024

$\sum \Delta P = -0.091$

Table D-4 FS = 2.0

I	α	L	H_L	H_R	V	W	U	$P_{i-1} - P_i$
1	-50.16	1.984	0	0	0	0.0083	0	-0.007
2	-53.05	3.813	0	0	0	0.0625	0	-0.047
3	9.50	9.270	0.7289	0.0453	0	1.7864	0.6925	-0.041
4	36.95	2.538	0	0	0	0.0479	0	0.064
5	33.62	2.752	0	0	0	0.0189	0	0.028

$\sum \Delta P = -0.003$

The safety factor for sliding equilibrium of the five-wedge system is determined from:

$$\sum_{i=1}^{5} (P_{i-1} - P_i) = \Delta P_R \quad \Delta P_R = 0$$ which gives the safety factor for equilibrium.

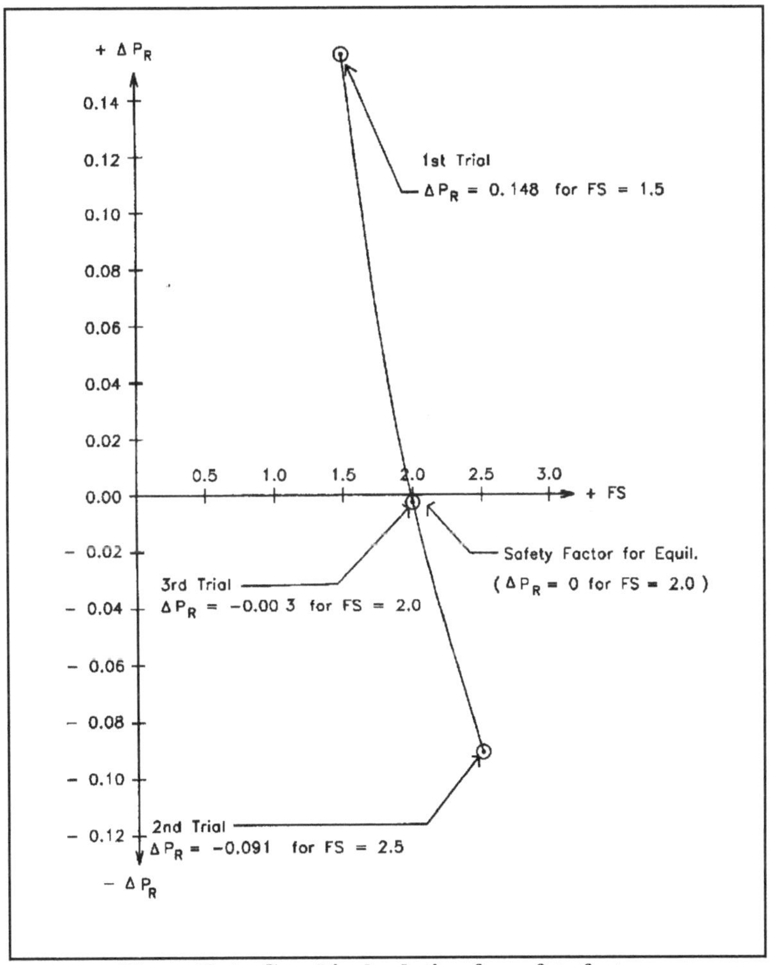

Figure D-4 Graphical solution for safety factor

D-3. Example D3, Lock Monolith on Rock.

Perform a complete stability analysis for the lock chamber wall monolith shown in Figure D-5. The structure classification is *normal*, the loading is a *usual* condition, and site information is *ordinary*. Assume that a vertical crack exists from the top of the rock to the base of the wall on the driving side. It is also assumed that movement of the wall will not be sufficient for any significant rock force to develop on the resisting side. From Chapter 3, the minimum factor of safety against sliding for a *usual* load condition, a normal structure classification, and *ordinary* site information is 1.50.

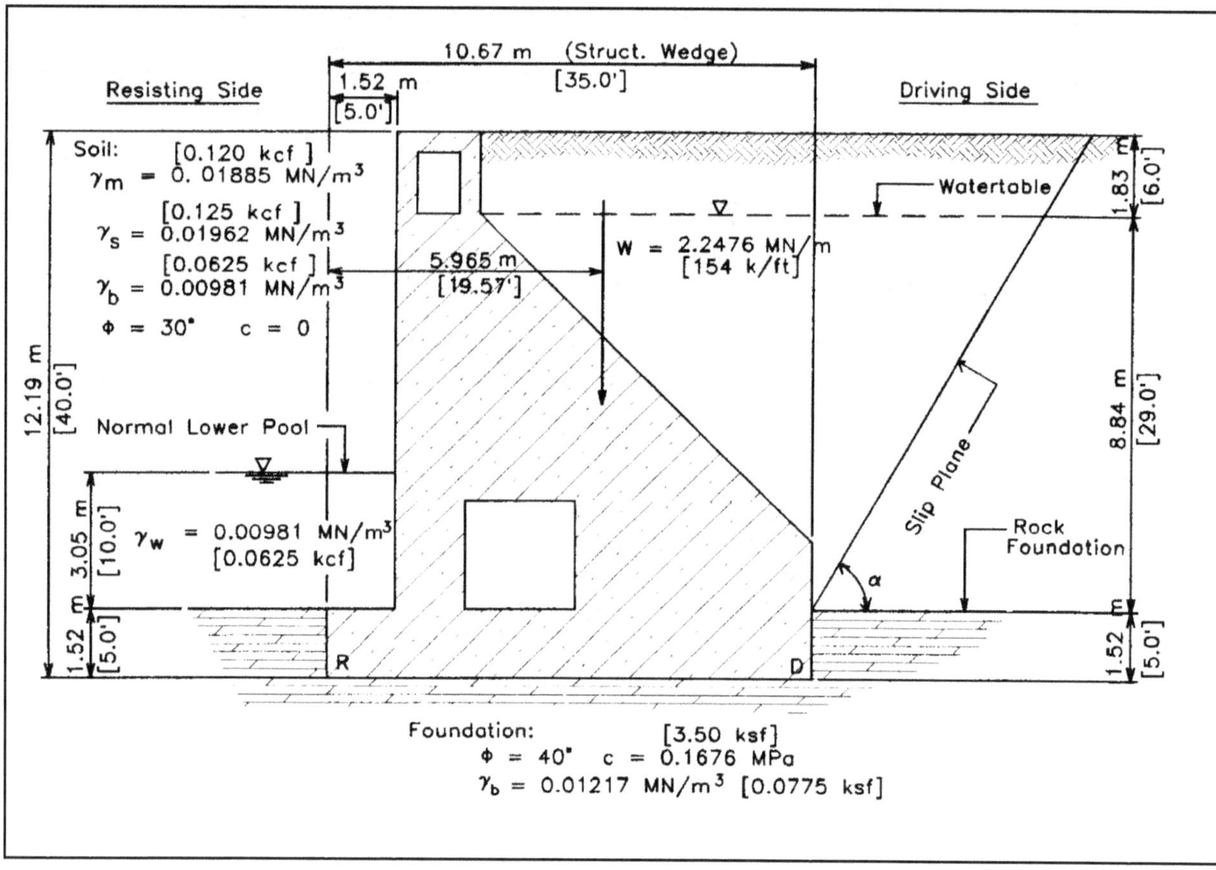

Figure D-3

a. Driving Side, Lateral Water Force

$$p_s = K\,\gamma_m (h - h_s) \quad p_s = 0.4714 \times 0.01885 \times 1.83 = 0.01626\ MN/m\ (0.3395\ \text{ksf})$$

$$P_w = \frac{1}{2}\gamma_w h_s^2 = \frac{1}{2}(0.00981)(\ 10.36\ )^2 = 0.5265\ MN/m\ [36.13\ \text{k/ft}]$$

b. Driving Side, Soil

$$\phi_d = \tan^{-1}\left(\frac{\tan 30^\circ}{1{,}50}\right) = 21.05^0,\ \ K = \tan^2\left(45^\circ - \frac{\phi_d}{2}\right) = 0.4715$$

$$p = K[\gamma_m h - (\gamma_m - \gamma_b) h_s] \quad p = 0.4715[0.01885x10.67 - (0.01885 - 0.00981)884$$

$$p = 0.05715 MPa (1.1933 ksf)$$

$$P = \frac{1}{2}(0.01626)(1.83) + \frac{1}{2}(0.01626 + 0.05715)8.84 = 0.3394 MN/m \ [23.26 \text{ k/ft}]$$

c. *Resisting Side Lateral Water Force*

$$P_w = \frac{1}{2} x\ 0.0448\ x\ 4.57 = 0.1024\ MN/m\ (7.03\ k/ft)$$

d. Uplift

$$u_R = \gamma_w h_s = 0.00981\ x\ 4.57 = 0.0448, \quad u_D = 0.00981\ x\ 10.36 = 0.1016$$

$$U = 1/2\ (0.0448 + 0.1016)(10.67) = 0.7810\ MN/m\ (53.53\ k/ft)$$

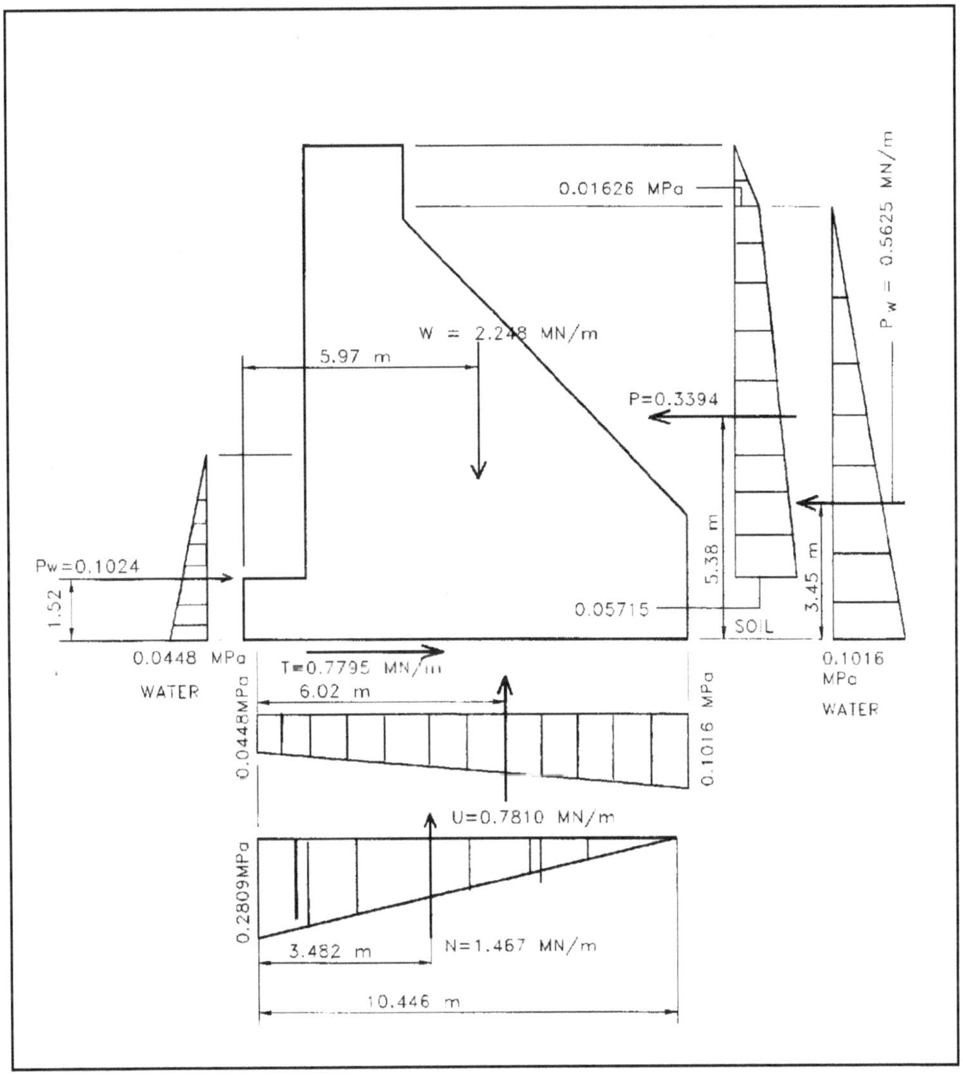

Figure D-6 Forces and pressures from single wedge analysis

e. Resultant Location for Single Wedge Sliding Analysis (see Figure D-6)

$$\Sigma M_R = 2.248x5.97 + 0.1024x1.52 - 0.3394x5.38 - 0.5265x3.45 - 0.7810x6.02 = 5.108$$

$$N = 2.248 - 0.781 = 1.467\ MN/m, \quad X_R = \frac{\Sigma M_R}{N} = 3.482\ m, \quad L = 3\,X_R = 10.446\ m < 10.67\ m$$

$$FS = \frac{N \tan\phi + cL}{T} = \frac{1.467x0.8391 + 0.1676x10.67}{0.4483} = 6.63 > FS_{sliding} = 1.50$$

f. *Results.* Factor of Safety Against Sliding Exceeds the Minimum Requirement. 97.9 % of base is in compression, but Resultant Location Criteria is assumed to be satisfied since any rock resistance has been neglected. Uplift will not be re-calculated

D-4. Example D4, Stability Benefits from Uplift Reduction

This example problem represents a non-overflow section (gravity dam), which is part of a hydroelectric power generating project. The non-overflow section was designed in the early 1900s under the assumption of full uplift acting over 50 percent of the base. Stability analyses performed under current stability criteria indicates the structure does not meet resultant location, and sliding factor of safety requirements. The dam section has a grouting and drainage system, and piezometer readings indicate uplift downstream of the drains is at or near tailwater. Maximum drain efficiency allowed for the design of new dams is 50 percent according to information provided in Appendix C. The stability will be evaluated using 50-percent drain efficiency. It will also be evaluated assuming 90-percent drain efficiency, which conservatively represents the degree of drain efficiency indicated by piezometer readings taken over the past 40 years.

a. Dam Configuration. A cross section through the highest section of the dam is shown in Figure D7. The pool conditions are indicated for the normal operating condition. A *usual* loading condition is used in establishing the required factors of safety for sliding and resultant location.

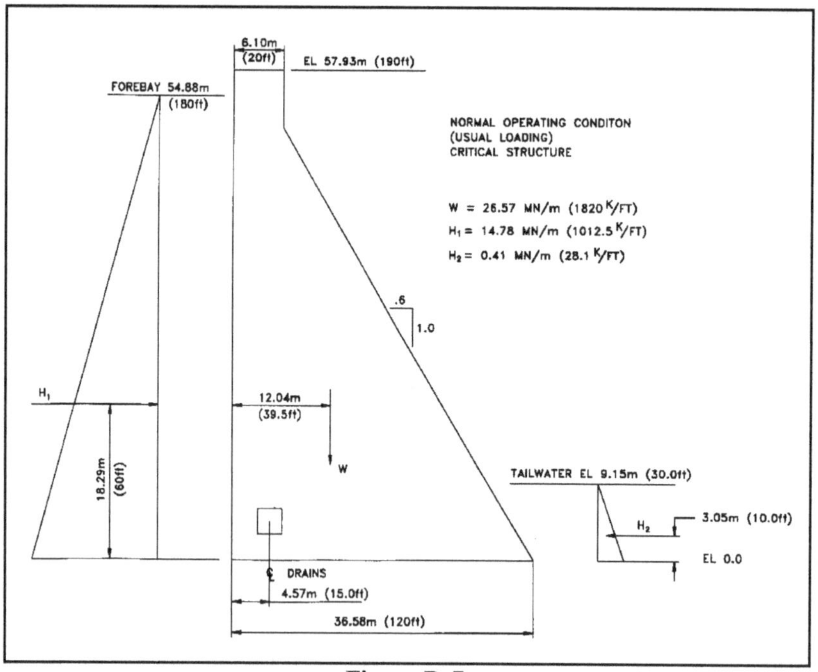

Figure D-7

b. Foundation strength parameters. Through explorations and testing it has been determined that the angle of internal friction for the foundation (ϕ) is 55-deg, and because the foundation rock is highly jointed and fractured, the cohesive strength (c) is assumed to be zero.

c. Loads. The forebay and tailrace pool elevations representing the usual loading condition are shown in Figure D-7. Table D-5 provides a summary of the loads and a summary of the moments of the loads about the heel of the dam.

d. Uplift. The uplift pressures and resultant locations for the 50-percent drain efficiency condition are shown in Figure D-8, and for the 90-percent drain efficiency in Figure D-9. The forebay pool, since this is a hydroelectric power generating facility, is usually kept at elevation 54.88 m (180 ft) to maximize power benefits. Piezometer measurements and records have been kept over the last 40 years. Many records are available for the forebay pool at elevation 54.88 m (180 ft) and with tailwater at a maximum elevation of 9.15 m (30.0 ft). The records indicate the uplift downstream of the drains remains at or near tailwater. The records also indicate that uplift pressures are not increasing with time. A comparison of the 90-percent drain efficiency condition with the actual measured uplift pressure condition for the normal operating forebay pool / maximum tailwater condition is shown in Figure D-10. It is apparent that the 90-percent drain efficiency condition conservatively represents the actual foundation uplift pressures.

Table D-5 Loads and Moments (without uplift)

Load Designation	Horizontal Load (MN)	Vertical Load (MN)	Moment Arm (meters)	Moment @ A (MN-meters)
H_1	14.78		18.29	270.33
H_2	-0.41		3.05	-1.25
W		26.57	12.04	319.90
SUM	14.37	26.57		588.98

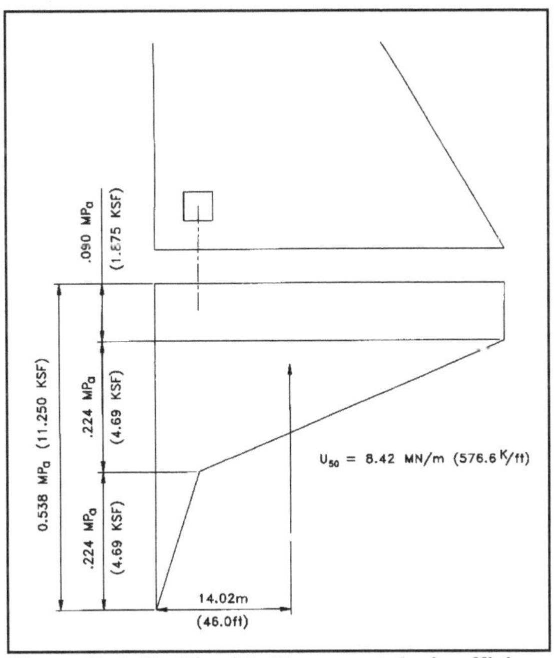

Figure D-8 Uplift with 50 percent drain efficiency (Resultant in middle 1/3 of base)

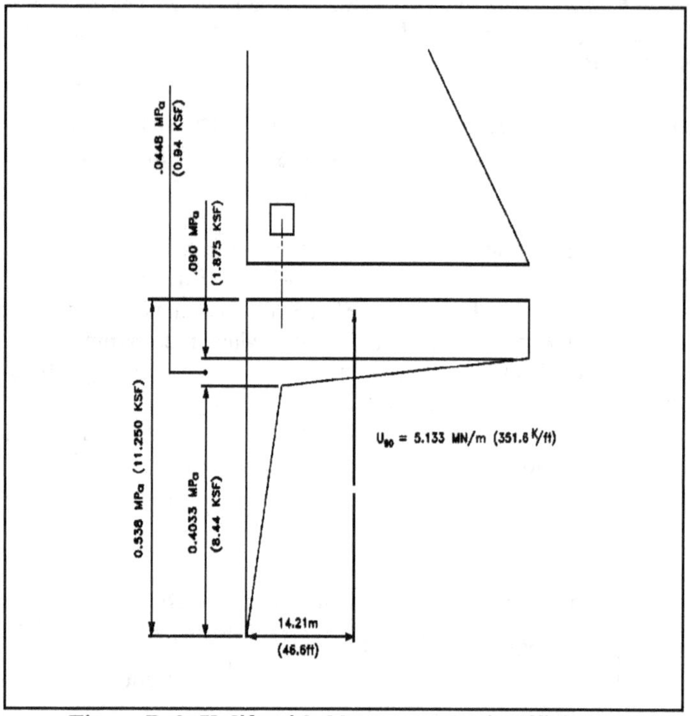

Figure D-9 Uplift with 90 percent drain efficiency

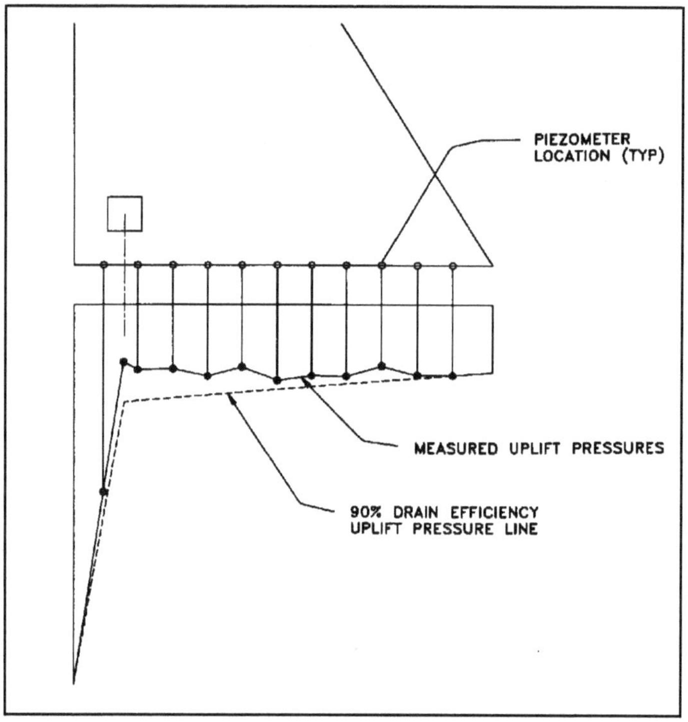

Figure D-10 Measured uplift versus 90% drain efficiency

e. Sliding Factors of Safety. The sliding factors of safety are calculated using the limit equilibrium method. For the single-plane failure surface condition with a horizontal failure plane, the factor of safety equation becomes:

$$FS = \frac{cA + (V \quad U\,)\tan\phi}{H} \text{ , and}$$

For the 50-percent drain efficiency condition:

$$FS = \frac{18.15 \tan 55}{14.37} = 1.80 < 2.00 \quad \text{N.G.}$$

For the 90-percent drain efficiency condition:

$$FS = \frac{21.44 \tan 55}{14.37} = 2.13 > 2.00 \quad \text{O.K.}$$

f. Conclusions. The use of a drainage efficiency of 90-percent provides the required factor of safety for sliding, whereas for the 50-percent drainage efficiency condition the structure fails to meet sliding requirements. This particular example problem illustrates the case where the use of existing uplift conditions can eliminate the need for a costly stability retrofit. A prudent designer, however, would make sure that the drains were inspected and cleaned on a regular basis, that there were sufficient piezometers throughout the base of the structure, and that these piezometers were read on a regular basis to remove all concerns over high uplift pressure leading to stability problems. There are many cases where drainage systems for dams have reduced uplift pressures to tailwater levels.

D-5. Example D5, Sliding of Gravity Dam With and Without Tensioned Structural Anchors

This example illustrates how sliding stability is enhanced by the use of tensioned structural anchors. The anchors accomplish stability by decreasing the component of the resultant force that acts parallel to the sliding plane and by increasing the component of the resultant force that acts normal to the sliding plane. Foundation strength parameters used in this example are $\phi = 40°$ and c=0. Force diagrams are shown in Figures D-11 and D-12.

<u>Without Anchors.</u>

$$N = W - U = 5.04 - 1.48 = 3.56 \text{ MN/m}(243.28 \text{ k/ft}); \; T = \Sigma H = 2.23 \text{ MN/m}(153.1 \text{ k/ft})$$

$$FS = \frac{N \tan \Phi}{T} = \frac{(3.56)(0.8391)}{2.23} = 1.34 < 2.0 \text{ N.G.}$$

<u>With Anchors.</u>

$$N = W + F \sin 45° - U = 5.04 + (1.21)(0.707) - 1.28 = 4.61 \text{ MN/m} (315.9 \text{ k/ft})$$

$$T = 2.23 - F \cos 45° = 2.23 + (1.21)(0.707) = 1.38 \text{ MN/m} (94.6 \text{ k/ft})$$

$$FS = \frac{N \tan\phi}{\text{T}} = \frac{4.61 \text{ x } 0.8391}{1.38} = 2.80 > 2.0 \text{ O.K.}$$

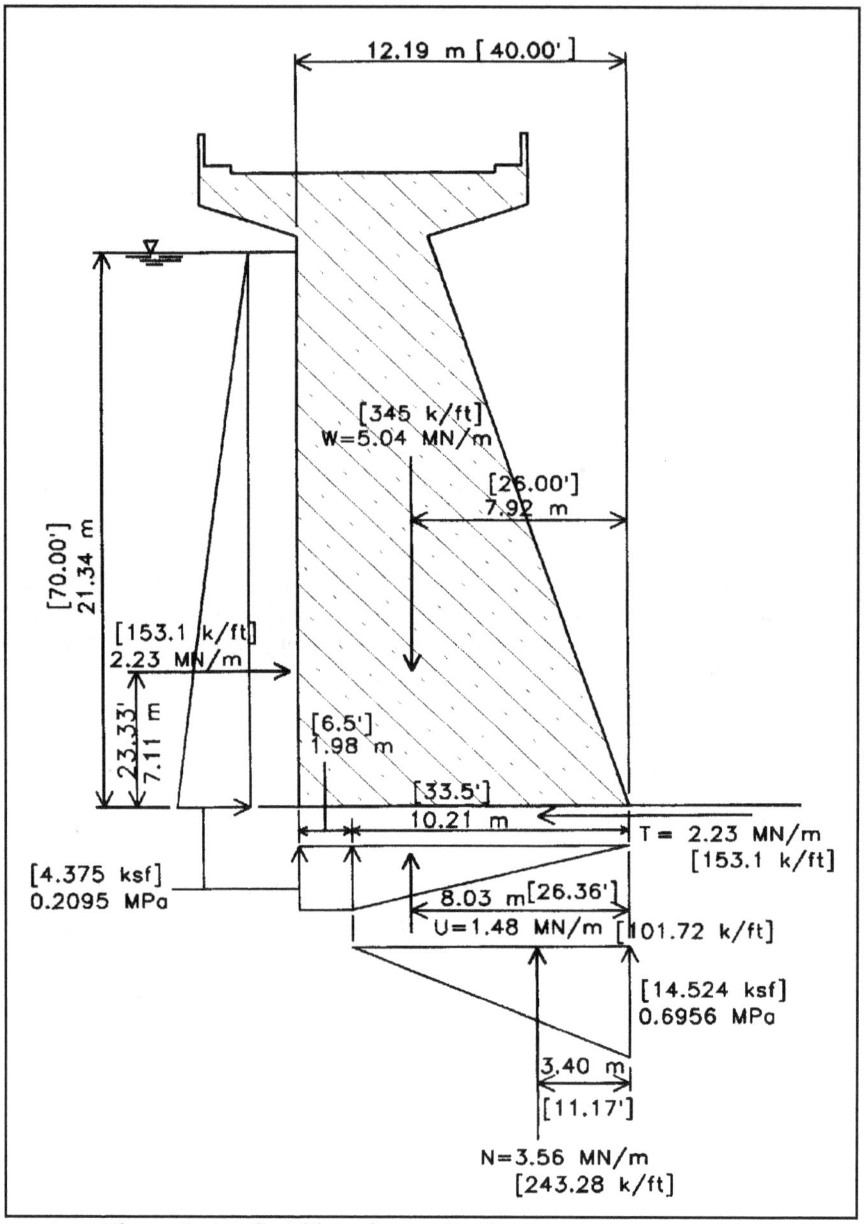

Figure D-11. Stability without tensioned structural anchors

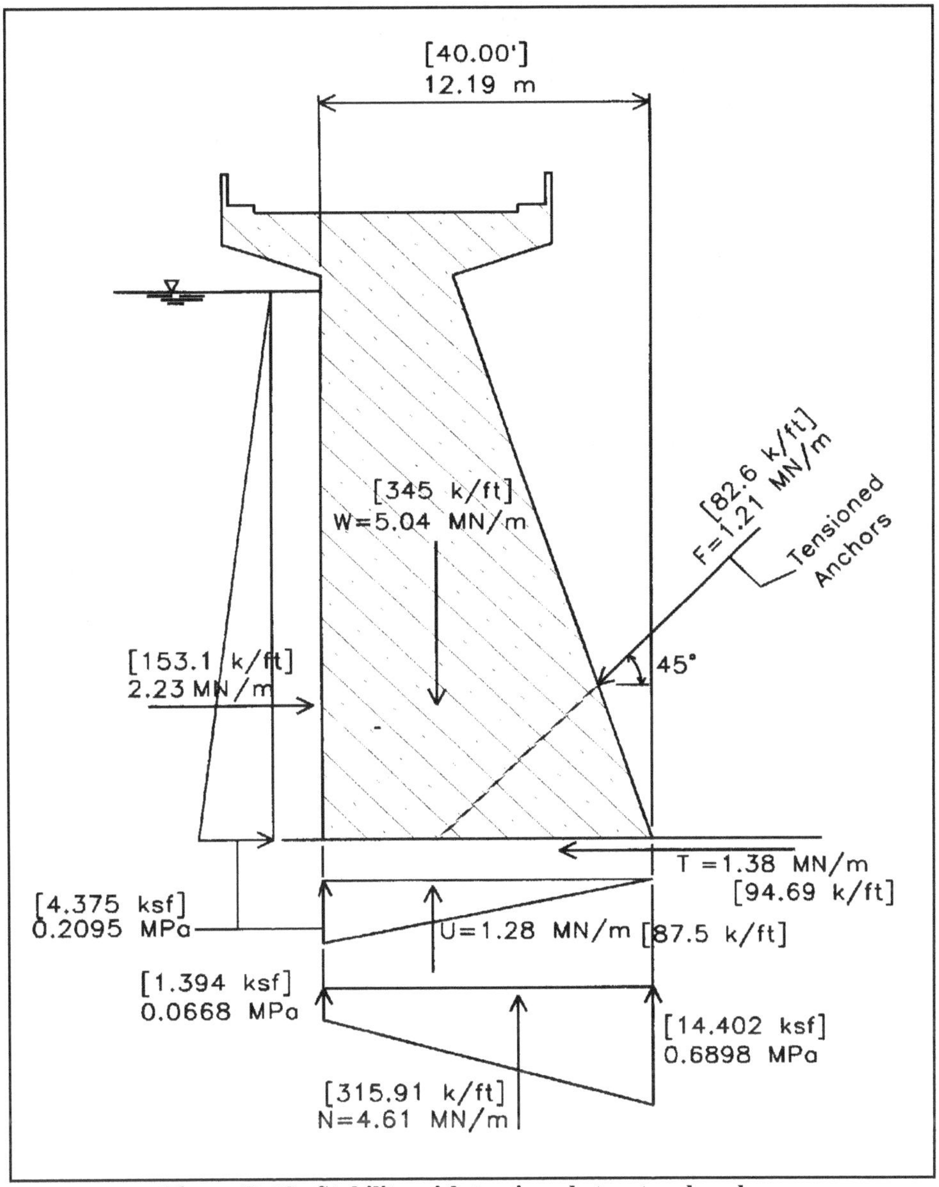

Figure D-12. Stability with tensioned structural anchors

D-6. Example D6, Obtaining Shear Strength Parameters for Use in Sliding-Stability Analyses

The limit equilibrium approach, used for determining a sliding factor of safety, will only provide meaningful results when there is a high degree of confidence that the shear strength parameters used are a conservative representation of actual foundation strengths. The selection of the shear strength parameters involves a great deal of uncertainty. Shear strength parameters can be determined through laboratory and in situ testing, by empirical methods, or by judgment based on a knowledge of rock mechanics and geological conditions. The type of methods and testing used depends on whether the sliding failure plane consists of intact rock, jointed rock, or sheared rock zones containing a weak filler material. Although laboratory testing is the most common way of determining shear strength parameters, the strength of laboratory samples may or may not be indicative of the strength of the prototype. In general, due to size effects, the laboratory samples will provide strengths greater than that of the prototype.

a. Stress / Strain Relationships. There is no consideration of strain in the limit equilibrium method. For foundations that contain more than one material, or are made up of a combination of intact rock and jointed or sheared rock, the limit equilibrium method is only valid when the materials behave in a typical elastic-plastic manner. This means that once the intact rock reaches its peak shear strength, the strains necessary to reach the peak strengths of the jointed and sheared materials can be obtained without a loss of strength through strain softening in the intact rock. For strain softening situations, some judgments must be made with respect to the selection of strength parameters which represent failure strain conditions.

b. Zintel Canyon Project – Foundation Materials. The Foundation and Seepage Analysis Report for the Zintel Canyon Project is used to illustrate the foundation strength determination process for a dam constructed on jointed basalt with various degrees of weathering. The descriptions of the various materials, which make up the foundation, and the laboratory test results, are described below. This information was taken directly from the referenced report.

(1) General. The materials mapped in the foundation of Zintel Canyon Dam consist of a basalt bedrock. The basalt is moderately to intensely weathered and has been subjected to heat alteration and shearing. Shears were mapped in the left abutment, stilling basin, and key trench.

(2) Slightly to moderately weathered basalt. Approximately 20 percent of the total foundation area of Zintel Canyon Dam consists of slightly weathered to moderately weathered basalt. The closely jointed rock resulted in a rough foundation surface with tightly interlocked angular pieces of basalt extending 152.4 mm (6-in.) or more above the lowest portions of the cleaned foundation surface. Drill hole DH-6RD was completed in the slightly to moderately weathered basalt in the right abutment area. Core samples recovered from this hole consisted of closely jointed, hard basalt. The bi-axial shear tests results on the core samples are shown in Figure D-13.

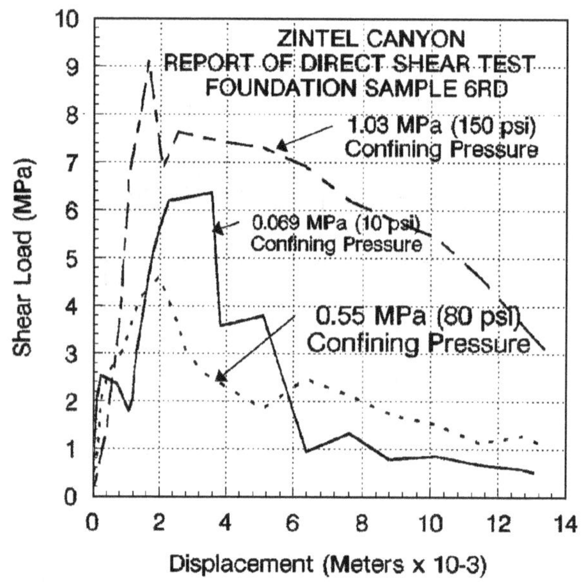

Figure D-13. Direct shear test for slightly to moderately weathered basalt

(3). Moderately to intensely weathered basalt. Approximately 55 percent of the total foundation of Zintel Canyon Dam consists of moderately weathered to intensely weathered basalt. The higher degree of weathering, and very close fracturing in these areas of the foundation, resulted in angular pieces of the basalt being easily loosened and requiring removal during foundation cleaning. The more weathered condition of the basalt resulted in a somewhat more irregular foundation surface than in the slightly to moderately weathered areas. Drill hole DH-1SB was completed in a typical area of the moderately weathered to intensely weathered basalt. Core samples from this hole consisted of very closely fractured and jointed weathered basalt. The bi-axial shear test results on the core samples are shown in Figure D-14.

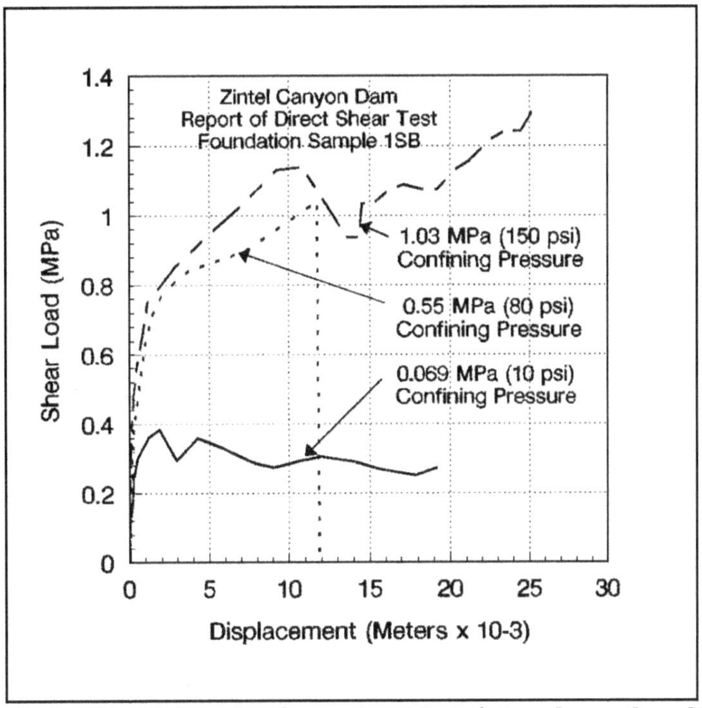

Figure D-14. Direct shear test for moderately to intensely weathered basalt

(4) Intensely weathered / altered basalt. The intensely weathered / altered basalt comprises approximately 20 percent of the total area of the dam foundation. The condition of the rock varies throughout the mapped areas. Commonly, the weathering and alteration resulted in a foundation surface characterized by gravel-sized loosely interlocked angular pieces of basalt extending less than 76.2 mm (3-in.) above the lowest portions of the cleaned foundation. However, some limited areas of the intensely weathered and highly altered basalt are characterized by fragments within a matrix of soil-like highly altered and decomposed basalt. The highly altered areas of the foundation were excavated somewhat deeper than the surrounding areas. However, there was no indication the material was improving with depth, or that by excavating deeper the material would improve. Drill hole DH-1SB, completed in the right side of the stilling basin, encountered the intensely weathered and altered basalt. Core samples from this hole were disturbed during the drilling operation and were primarily recovered as angular basalt fragments. On the basis of the recovered core samples and foundation observations, the intensely weathered and altered areas of the foundation generally consist of a poorer-quality rock than in the moderately to intensely weathered areas, but better-quality rock than in the sheared areas where a large percentage of the cored material was recovered as basalt fragments in a matrix on clay or silt. Therefore, the shear strength of the basalt in the intensely weathered and altered areas of the foundation is considered to be somewhat higher than in the sheared areas.

(5) Shear zones. Shears make up approximately 5 percent of the total foundation of Zintel Canyon Dam. Shears were mapped in the left abutment, stilling basin, and key trench. The shears are predominantly high angle and are mostly orientated from North 5-deg West to North 46-deg East. With this orientation, the shears do not appear to threaten the safety of the structure. There is no indication that the shears would intercept one another at shallow depths, creating the potential for a block failure, or that they form planes of significant weakness in the foundation. Also, due to the infilling of clay and/or silt within the shears, the shear planes are not expected to provide pathways for seepage under the dam. The shears primarily consist of sand - to-gravel-sized fragments of basalt in a matrix of clay or silt. The sheared foundation surface areas are somewhat smoother than other areas of the foundation. Drill hole DH-7LD, located near the base of the left abutment was completed in the area of shear S-9. The core samples from this hole consisted of angular to sub-angular fragments of basalt in a clay matrix. The bi-axial shear test results on the core samples are shown in Figure D-15.

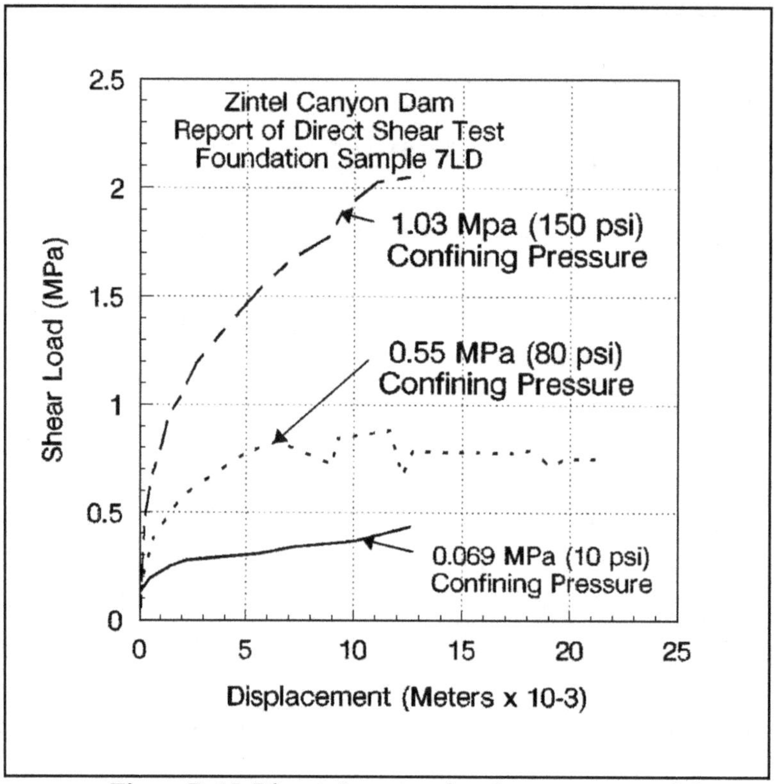

Figure D-15 Direct-shear test for shear zone material

c. Foundation samples and laboratory test results.. Foundation core samples from drill holes, DH-1SB, DH-6RD and DH-7LD were prepared for bi-axial direct-shear strength tests. Sample selections were dependent on core condition (absence of drilling and handling disturbance). In addition, the samples were selected so that the typical best, average, and poorest quality foundation rock encountered would be tested..

(1) Bi-axial shear sample preparation and test procedure. Bi-axial direct-shear strength tests were performed at confining pressures of 1.03, 0.55, and 0.069 MPa (10, 80, and 150 psi.). A test section of 0.762 to 0.915 m (2.5 to 3.0 ft) per hole (two sections in DH-7LD) was selected. In general, each bi-axial shear strength specimen was saw cut to a nominal 0.2033 m (8-in.) length, placed in steel shear blocks, plastered in place and tested. Rock cores from drill holes DH-1SB and DH-7LD could not be handled without breaking apart, so specimens from these holes were prepared by wire-banding the original PVC encasement around the core and saw-cutting through both the PVC and the rock core. Specimens from drill hole DH-1SB required a wax coating in order to keep the samples intact after removal from the PVC encasement prior to placing in the shear blocks. Bi-axial testing was determined to be the most practical method for testing this material. Tri-axial testing was not considered appropriate. The required sample preparation and significant handling requirements made it highly unlikely that an undisturbed rock sample could be situated in the available tri-axial chamber. It is recognized that tri-axial testing of poorer-quality rock would yield improved shear performance, if an undisturbed sample could have been prepared.

(2) Test results. Load-displacement curves from the bi-axial shear tests are shown in Figures D-13 through D-15, and the results in terms of shear strength versus confining pressure are plotted in Figure D-16. The rock samples describes as hard basalt, minor fractures (DH-6RD), representing approximately 20 percent of the foundation rock yielded a cohesive value of 1.6552 MPa (240 psi) and angle of internal friction of 82 deg. The rock samples described as severely fractured, weathered basalt, minor infilling (DH-1SB), representing approximately 55 percent of the foundation rock yielded a cohesion value of 0.414 MPa (60 psi) and an angle of internal friction of 39 deg. The rock samples described as weathered sheared basalt, fully infilled with altered clay / sand (DH-7LD), representing 5 percent of the foundation rock yielded a cohesion value of 0.0897 MPa (13 psi) and an angle of internal friction of 56 deg. The test results from several of the shear tests were not peak values of shear load. They were actually the value of shear when the test was concluded.

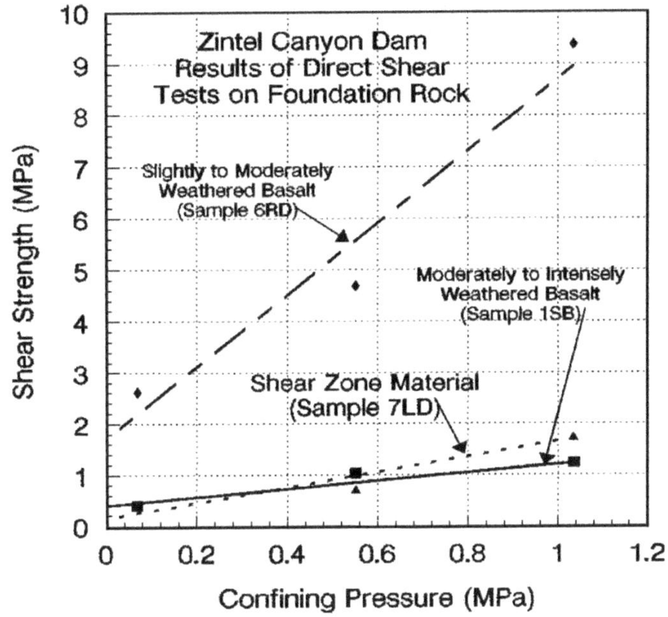

Figure D-16 Shear strength versus confining pressure

d. Observations. The example problem is used to illustrate the difficulties in establishing foundation strength parameters for foundation rock that is jointed and sheared. Intact foundation rock generally poses no problems in the selection of strength parameters, because even with extremely conservative values the foundation has no difficulty in carrying the loads imposed on it by the structure. As it can be seen from the Zintel Canyon Dam direct-shear load / displacement curves, the foundation often exhibits strain-softening characteristics rather than the elastic-plastic characteristics assumed in the limit-equilibrium method. Therefore, the stress/strain relationships of the various foundation materials play an important part in the selection of strength parameters to be used in the sliding stability analysis. This example also demonstrates how erratic the tests results can be and how difficult it can be to establish a single set of strength parameters from the test results. For Zintel Canyon Dam, conservative lower-bound shear-strength values representing the sheared basalt material provided a sliding factor of safety greater than required. Because lower-bound shear-strength parameters were used in the sliding stability analysis, a normal site information factor was selected to establish the required factor of safety.

D-7. Example D7, Spillway Slab With And Without Untensioned Structural Anchors.

This example compares the spillway slab thicknesses required to withstand flotation, with and without the use of untensioned structural anchors. The analysis is for a condition referred to as *rapid drawdown*, which occurs when the spillway gates are closed suddenly. This results in the entrapment of uplift pressure under the slab for a short period of time while the actual water surface is near or at the top surface of the slab. In this example, the entrapped uplift head is assumed to be 3.049m + t ([10 ft + t]). The plan and elevation of the spillway slab panel used in this example are shown in Figure D-17. It is assumed for this example that the return period for the loading condition results in an *unusual* loading condition classification, *normal* importance classification, and *normal* site information.

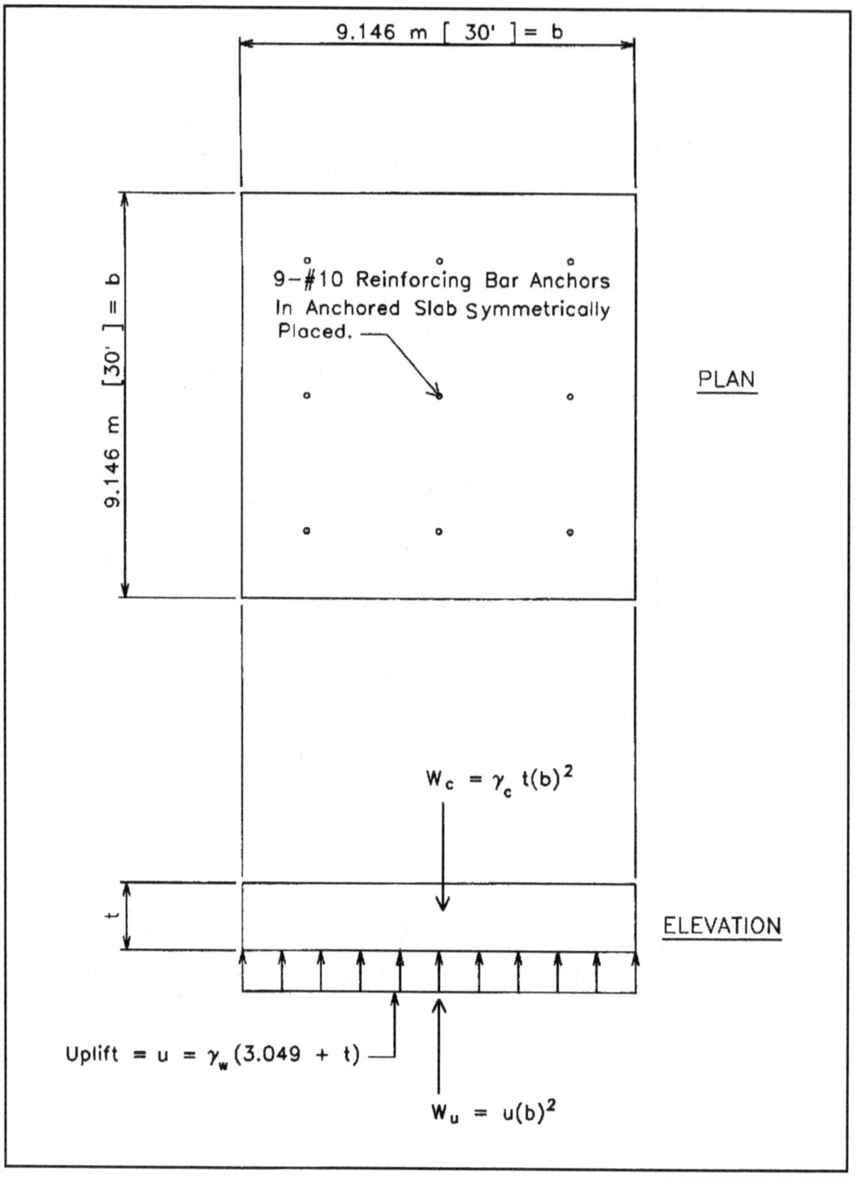

Figure D-17 Anchored spillway slab

a. Thickness with anchors.

$$W_u = \gamma_w (3.049 + t)\, b^2 = 0.00982\,(3.049 + t)(9.146\,)^2 = 2.50456 + 0.82144\, t$$

$$W_c = \gamma_c b^2 t = 1.97078\, t$$

The requirement of EM 1110-2-2104 for adequacy of reinforcing bar is $\phi R \geq U$ where $\phi = 0.9$ for tension. Thus:

$$U = 0.75[1.3(1.7\, W_u - 1.4\, W_c)], \quad R = N A_s\, f_y$$

$$U = 0.75\,(1.3\,[1.7\,(2.50456 + 0.82144\,t) - 1.4\,(1.97078\,t)])$$

$$U = 4.15131 - 1.32858\,t$$

$$\phi R = 0.9 \times 9 \times 0.0008198 \times 413.793 = 2.74774\ MN\ (617.7k)$$

Setting ϕR equal to U and solving for t:

$$2.74774 = 4.15131 - 1.32858\,t$$

t = 1.0564 m (3.465 ft), Use 1.067 m (3.50 ft)

b. Thickness without anchors. From above

$$W_u = 2.50456 + 0.82144\,t, \quad W_c = 1.97078\,t$$

From Chapter 3 for an *unusual* load condition, a *normal* structure, and *normal* site information the required factor of safety against flotation is FS = 1.20. Setting W_c equal to FS(W_u) and solving for t:

$$1.97078\,t = 1.20\,(2.50456 + 0.82144\,t)$$

$$0.98505\,t = 3.00547, \quad t = 3.051\,m\,(10.00\ ft)$$

c. Conclusion. This example illustrates the importance of untensioned structural anchors for slabs on grade subject to high uplift pressures. It is obvious from inspection that the anchored slab would be more economical, i.e., 1.067 m (3.5 ft) versus 3.051 m (10.0 ft) thick slab.

D-8. Example D8, Critical Slip Plane Angle (α) For An Earth Wedge With A Broken Top Surface

This driving-side wedge has a broken top surface, the water table lies within the wedge, and the soil possesses internal friction only, as shown in Figure D-18.

First, it is assumed that the top surface of the wedge is horizontal and lies along line ABC. Then the weight of the area BCD will be taken as an analogous negative strip surcharge (V). Where

$$V = -\frac{1}{2}(0.01886)\,(7.5)\,(3) = -0.212175\ MN/m\ [-14.53\ \text{k/ft}]$$

Since the water table lies within the wedge, the average effective unit weight of soil will be used in the calculations for the critical-slip plane angle.

$$\gamma = \gamma_{avg} = \gamma_m - \frac{(\gamma_m - \gamma_b)\,h_s^2}{h^2}$$

$$\gamma = 0.01886 - \frac{0.00904\,(7)^2}{(18)^2} = 0.017493\ MN/m^3\ [0.11132\ \text{kcf}]$$

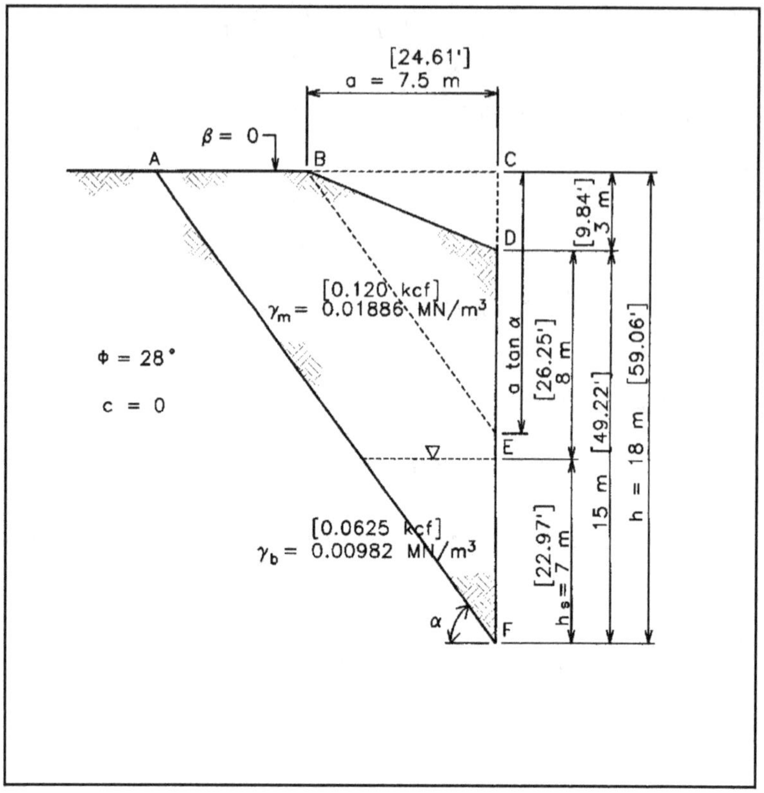

Figure D-18

Using the equations in Chapter 5, calculate the value of the critical-slip plane angle.

$$r = 1 - \tan\beta \tan\phi = 1 - 0 \, x \, 0.531709 = 1.00$$

$$s = \tan\phi + \tan\beta = 0.531709 + 0 = 0.531709$$

$$t = \tan\phi - \tan\beta \tan^2\phi = 0.531709 - 0\,(0.531709\,)^2 = 0.531709$$

$$\alpha = \tan^{-1}\left(\frac{C_1 + \sqrt{C_1^2 + C_2^2}}{2}\right) = 55.8637^0$$

$$C_1 = \frac{2\,s\tan\phi}{A} = \frac{2\,x\,0.531709\,x\,0.531709}{0.627747} = 0.900727$$

$$C_2 = \frac{t}{A} = \frac{0.531709}{0.627747} = 0.847012$$

$$A = s - \left(\frac{2V}{\gamma h^2}\right)(1 - \tan^2\phi) = 0.531709 - \left(\frac{-2\,x\,0.212175}{0.017493\,(18\,)^2}\right)(1.2182714) = 0.627747$$

D-9. Example D9, Critical Slip Plane Angle (α) with Cohesive Soil

The wedge in this example is the same as the wedge in Example D8, except that the soil possesses cohesive strength as well as internal friction, as shown in Figure D-19.

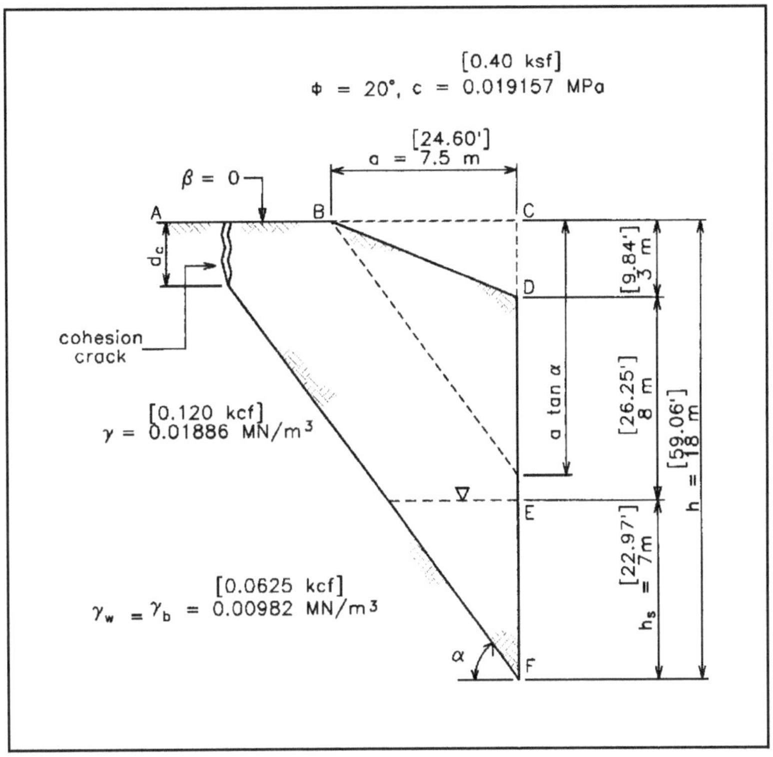

Figure D-19

Again, it is assumed that the top surface of the wedge is horizontal and lies along the line ABC. The weight of the area BCD will be taken as an analogous negative surcharge load (V).

Where

$$V = -\frac{1}{2}(0.01886)\ (7.5)\ (3) = -\ 0.212175\ MN/m\ [\ 14.53\ \text{k/ft}\ ]$$

Since the water table lies within the wedge, the average effective unit weight of soil will be used in determining the critical slip plane angle for the wedge.

$$\gamma = \gamma_{avg} = \gamma_m - \frac{(\gamma_m - \gamma_b)\, h_s^2}{(h^2 - d_c^2)}$$

Since the depth of cohesive cracking (d_c) is not known initially, it must be estimated at the start and the equations in Chapter 5 used in an iterative fashion until the critical-slip plane angle (α) is found that produces a calculated value for crack depth that is equal to the estimated value.

$$r = 1 - \tan\phi \tan\beta = 1.00$$

$$s = \tan\phi + \tan\beta = 0.36397$$

$$t = \tan\phi - \tan\beta \tan^2\phi = 0.36397$$

<u>1st Trial:</u> estimate dc = 3 m

$$\gamma = \gamma_{avg} = 0.01886 - \frac{0.00904\,(7)^2}{(18)^2 - (3)^2} = 0.017454\ MN/m^3\ [0.11107\ \text{k/ft}^3]$$

$$\gamma\,(h^2 - d_c^2) = 5.49801, \quad \gamma\,(h + d_c) = 0.366534$$

$$A = s - \left(\frac{2V}{\gamma\,(h^2 - d_c^2)}\right)(1 + \tan^2\phi) + \left(\frac{2c}{\gamma\,(h + d_c)}\right) r$$

$$A = 0.36397 - \left[\frac{2\,(-0.212175)}{5.49801}\right](1.132474) + \left(\frac{2 \times 0.019157}{0.366534}\right)(1.00) = 0.555908$$

$$C_1 = \frac{2\,s\tan\phi + \left(\frac{4c}{\gamma\,(h + d_c)}\right) s}{A}$$

$$C_1 = \frac{2\,(0.36397)\,(0.36397) + \left(\frac{4 \times 0.019157}{0.366534}\right)(0.36397)}{0.555908} = 0.613483$$

$$C_2 = \frac{t + \left(\frac{2c}{\gamma\,(h + d_c)}\right) r}{A}$$

$$C_2 = \frac{0.36397 + \left(\frac{2 \times 0.019157}{0.366534}\right)(1.00)}{0.555908} = 0.842766$$

$$\alpha = \tan^{-1}\left(\frac{C_1 + \sqrt{C_1^2 + 4C_2}}{2}\right) = 51.885°$$

Calculate d_c and compare to the initial estimated value.

$$d_c = \frac{2K_c c}{K\gamma_m}, \quad K = \frac{1 - \tan\phi \cot\alpha}{1 + (\tan\phi)\tan\alpha} = 0.488038$$

$$K_c = \frac{1}{2\sin\alpha\cos\alpha\,[\,r + s\tan\alpha\,]} = 0.703303, \; d_c = \frac{2 \times 0.703303 \times 0.019157}{0.488308 \times 0.01886} = 2.93\,m$$

This is so close to the estimated initial value that further iteration is not necessary, and <u>$\alpha = 51.9°$</u>.

Appendix E
Wedge Equations

1. General Wedge Equation.

a. Nomenclature

α_i = angle between failure plane and horizontal

$\varnothing_i$ = internal friction angle of material in *ith* wedge. Earth and rock wedges only.

c_i = cohesive strength of material in *ith* wedge. Earth and rock wedges only.

W_i = weight of material in *ith* wedge.

V_i = surcharge load acting on *ith* wedge.

H_{Li} = horizontal force on *ith* wedge, acting to the right.

H_{Ri} = horizontal force on *ith* wedge, acting to left.

U_i = uplift (water) load acting normal to failure plane.

N_i = force acting normal to failure plane of *ith* wedge.

T_i = shear force acting parallel to failure plane of *ith* wedge.

P_i = horizontal force due to the *ith* wedge.

L_i = length of wedge base.

b. Sign Convention. The x and y axes are to the right and upward, respectively. The t and n axes are tangent and normal to the failure plane. Where the failure plane is not horizontal, the angle from the x-axis to the t -axis is determined by the right hand rule.

c. Equilibrium equations.

$\Sigma F_n = 0$

$$0 = N_i + U_i - W_i \cos\alpha_i - V_i \cos\alpha_i - H_{Li} \sin\alpha_i + H_{Ri} \sin\alpha_i - (P_{i-1} - P_i) \sin\alpha_i$$

$$N_i = (W_i + V_i) \cos\alpha_i - U_i + (H_{Li} - H_{Ri}) \sin\alpha_i + (P_{i-1} - P_i) \sin\alpha_i$$

$\Sigma F_t = 0$

$$0 = -T_i - W_i \sin\alpha_i - V_i \sin\alpha_i + H_{Li} \cos\alpha_i - H_{Ri} \cos\alpha_i + (P_{i-1} - P_i) \cos\alpha_i$$

$$T_i = (H_{Li} - H_{Ri}) \cos\alpha_i - (W_i + V_i) \sin\alpha_i + (P_{i-1} - P_i) \cos\alpha_i$$

d. Mohr-Coulomb failure criterion.

$$T_F = N_i \tan\phi_i + c_i L_i$$

e. Sliding factor of safety definition.

$$FS_i = \frac{T_F}{T_i} = \frac{N_i \tan\phi_i + c_i L_i}{T_i}$$

f. Governing wedge equation.

$$FS_i = \frac{\left[(W_i + V_i)\cos\alpha_i - U_i + (H_{Li} - H_{Ri})\sin\alpha_i + (P_{i\text{-}1} - P_i)\sin\alpha_i\right]\tan\phi_i + c_i L_i}{(H_{Li} - H_{Ri})\cos\alpha_i - (W_i + V_i)\sin\alpha_i + (P_{i\text{-}1} - P_i)\cos\alpha_i} \qquad \text{(E-1)}$$

$$(P_{i\text{-}1} - P_i) = \frac{\left[(W_i + V_i)\cos\alpha_i - U_i + (H_{Li} - H_{Ri})\sin\alpha_i\right]\dfrac{\tan\phi_i}{FS_i} - (H_{Li} - H_{Ri})\cos\alpha_i + (W_i + V_i)\sin\alpha_i + \dfrac{c_i L_i}{FS_i}}{\cos\alpha_i - \dfrac{\tan\phi_i \sin\alpha_i}{FS_i}}$$

A negative value of the difference $(P_{i\text{-}1} - P_i)$ indicates that the applied forces acting on the ith wedge exceed the forces resisting sliding along the base of the wedge. A positive value of the difference $(P_{i\text{-}1} - P_i)$ indicates that the applied forces acting on the ith wedge are less than the forces resisting sliding along the base of that wedge. The governing equation for $(P_{i\text{-}1} - P_i)$ applies to the individual wedges. For the system of wedges to act as an integral failure mechanism, the safety factors of all wedges must be identical.

$$FS_1 = FS_2 = \ldots\ldots FS_{i\text{-}1} = FS_i = FS_{i+1} = \ldots\ldots FS_N$$

where N = the number of wedges in the failure mechanism.

The actual safety factor for sliding equilibrium is determined by satisfying overall horizontal equilibrium ($\Sigma F_H = 0$) for the entire system of wedges.

$$\sum_{i=1}^{N} (P_{i\text{-}1} - P_i) = 0 \qquad \text{and} \qquad P_0 \equiv 0 \qquad P_N \equiv 0$$

Usually, an iterative solution process is used to determine the actual safety factor for sliding equilibrium. An example of a typical static loading condition analysis for a multiple-wedge system is presented in Example D2 of Appendix D. Note that if $\Sigma F_H < 0$, the factor of safety is less than the trial factor of safety, and if $\Sigma F_H > 0$, the factor of safety is greater than the trial factor of safety.

2. Critical Slip Angle for Driving-Side Wedge with Wall Friction on the Vertical Face

This section illustrates derivation of the critical slip angle for the limit equilibrium condition. Figure E-1 shows the forces acting on a driving-side wedge.

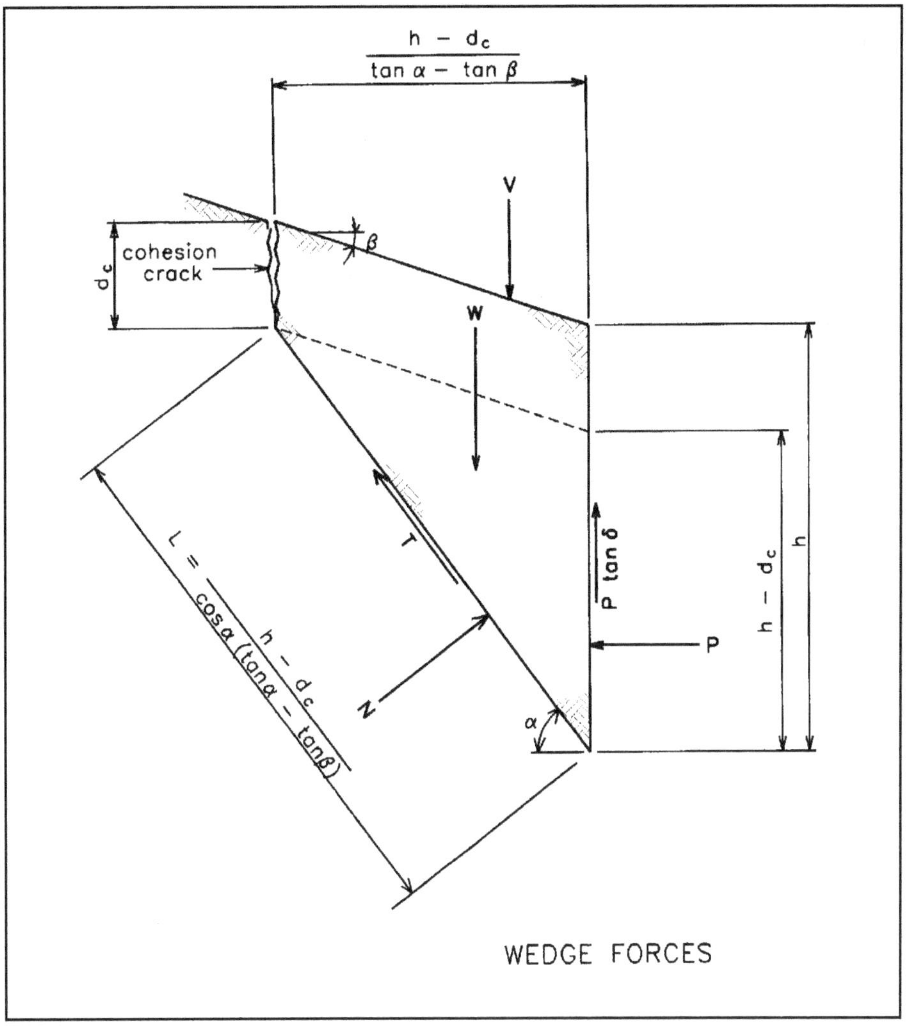

Figure E-1 Forces acting on wedge

For equilibrium to exist:

$$N = (W + V)\cos\alpha + P\sin\alpha - P\tan\delta\cos\alpha$$

$$T = (W + V)\sin\alpha - P\cos\alpha - P\tan\delta\sin\alpha$$

$$T \text{ must also} = \mathrm{N}\tan\phi + \mathrm{cL}$$

$$\mathrm{T} = \mathrm{N}\tan\phi + \mathrm{cL} = (\mathrm{W} + \mathrm{V})\tan\phi\cos\alpha + \mathrm{P}\tan\phi\sin\alpha - \mathrm{P}\tan\delta\tan\phi\cos\alpha + \mathrm{cL}$$

Equating the two expressions for *T*, dividing them by cos α, and solving for *P*, the following is obtained:

$$P = \frac{(W + V)(\tan\alpha - \tan\phi) - \dfrac{cL}{\cos\alpha}}{(1 - \tan\delta\tan\phi) + (\tan\delta + \tan\phi)\tan\alpha}$$

where

W = weight of soil in wedge

V = strip surcharge

$$W = \frac{\gamma(h^2 - d_c^2)}{2(\tan\alpha - \tan\beta)}, \quad \gamma = \text{unit weight of soil}$$

$$L = \frac{h - d_c}{\cos\alpha\,(\tan\alpha - \tan\beta)}$$

substituting the above values for W and L into the equation for P:

$$P = \frac{\left[\frac{\gamma(h^2 - d_c^2)}{2(\tan\alpha - \tan\beta)} + V\right](\tan\alpha - \tan\phi) - \frac{c(h - d_c)}{\cos^2\alpha(\tan\alpha - \tan\beta)}}{(1 - \tan\delta\tan\phi) + (\tan\delta + \tan\phi)\tan\alpha}$$

$$Note\ that: \frac{1}{\cos^2\alpha} = \sec^2\alpha = 1 + \tan^2\alpha$$

divide both sides of this equation by:

$$\frac{\gamma(h^2 - d_c^2)}{2}, \quad \text{also substitute } 1 + \tan^2\alpha \text{ for } \frac{1}{\cos^2\alpha}$$

to obtain:

$$\frac{2P}{\gamma(h^2 - d_c^2)} = \frac{(\tan\alpha - \tan\phi) + \frac{2V}{\gamma(h^2 - d_c^2)}(\tan\alpha - \tan\beta)(\tan\alpha - \tan\phi) - \frac{2c(1 + \tan^2\alpha)}{\gamma(h + d_c)}}{(\tan\alpha - \tan\beta)[1 - \tan\delta\tan\phi + (\tan\delta + \tan\phi)\tan\alpha]} = \frac{m}{n}$$

combining terms the above equation becomes:

$$\frac{m}{n} = \frac{(\tan\alpha - \tan\phi) + \frac{2V}{\gamma(h^2 - d_c^2)}[\tan^2\alpha - (\tan\beta + \tan\phi)\tan\alpha + \tan\beta\tan\phi] - \frac{2c(1 + \tan^2\alpha)}{\gamma(h + d_c)}}{(\tan\delta + \tan\phi)\tan^2\alpha + [1 - \tan\delta\tan\phi - \tan\beta(\tan\delta + \tan\phi)]\tan\alpha - \tan\beta(1 - \tan\delta\tan\phi)}$$

The necessary condition for P to be either a maximum or a minimum is that the derivative of m/n, with respect to α, be equal to zero. The derivative of m/n is:

$$\frac{d(m/n)}{d\alpha} = \frac{n\left(\frac{dm}{d\alpha}\right) - m\left(\frac{dn}{d\alpha}\right)}{n^2} = 0$$

from this it can be seen that if both sides of the equation are multiplied by n^2 the maxima-minima condition becomes:

$$n\frac{dm}{d\alpha} - m\frac{dn}{d\alpha} = 0$$

After differentiating this equation, it will be found that all terms are multiplied by $\sec^2\alpha$ ($d\tan\alpha/d\alpha$). For simplification, both sides of the equation will be divided (at the start) by $\sec^2\alpha$ to eliminate it.

Since all of the terms in the numerator (*m*) are over a common denominator, differentiation will be done separately for each numerator term; they will be combined at the end to furnish a complete solution. The derivative $dn/d\alpha$ is constant for all terms.

$$\frac{dn}{d\alpha} = 2(\tan\delta + \tan\phi)\tan\alpha + 1 - \tan\delta\tan\phi - \tan\beta(\tan\delta + \tan\phi)$$

The "*W*" term ($\tan\alpha - \tan\phi$):

$$\frac{dm}{d\alpha} = 1$$

$$n\frac{dm}{d\alpha} = (\tan\delta + \tan\phi)\tan^2\alpha + (1 - \tan\delta\tan\phi)\tan\alpha - \tan\beta(\tan\delta + \tan\phi)\tan\alpha - \tan\beta(1 - \tan\delta\tan\phi)$$

$$m\frac{dn}{d\alpha} = 2(\tan\delta + \tan\phi)\tan^2\alpha + (1 - \tan\delta\tan\phi)\tan\alpha - \tan\beta(\tan\delta + \tan\phi)\tan\alpha$$
$$- 2\tan\phi(\tan\delta + \tan\phi)\tan\alpha - \tan\phi(1 - \tan\delta\tan\phi) + \tan\beta\tan\phi(\tan\delta + \tan\phi)$$

$$n\frac{dm}{d\alpha} - m\frac{dn}{d\alpha} = -(\tan\delta + \tan\phi)\tan^2\alpha + 2\tan\phi(\tan\delta + \tan\phi)\tan\alpha + \tan\phi - \tan\beta - (\tan\delta + \tan\beta)\tan^2\phi = 0$$

The "*V*" term:

$$\frac{dm}{d\alpha} = \left[\frac{4V}{\gamma(h^2 - d_c^2)}\right]\tan\alpha - \left[\frac{2V}{\gamma(h^2 - d_c^2)}\right](\tan\beta + \tan\phi)$$

$$n\frac{dm}{d\alpha} = \left[\frac{4V}{\gamma(h^2 - d_c^2)}\right](\tan\delta + \tan\phi)\tan^3\alpha + \left[\frac{4V}{\gamma(h^2 - d_c^2)}\right][1 - \tan\delta\tan\phi - \tan\beta(\tan\delta + \tan\phi)]\tan^2\alpha$$
$$- \left[\frac{4V}{\gamma(h^2 - d_c^2)}\right]\tan\beta(1 - \tan\delta\tan\phi)\tan\alpha - \left[\frac{2V}{\gamma(h^2 - d_c^2)}\right](\tan\beta + \tan\phi)(\tan\delta + \tan\phi)\tan^2\alpha$$
$$- \left[\frac{2V}{\gamma(h^2 - d_c^2)}\right](\tan\beta + \tan\phi)[1 - \tan\delta\tan\phi - \tan\beta(\tan\delta + \tan\phi)]\tan\alpha$$
$$+ \left[\frac{2V}{\gamma(h^2 - d_c^2)}\right]\tan\beta(\tan\beta + \tan\phi)(1 - \tan\delta\tan\phi)$$

$$m\frac{dn}{d\alpha}=\left[\frac{4V}{\gamma(h^2-d_c^2)}\right][(\tan\delta+\tan\phi)\tan^3\alpha-(\tan\delta+\tan\phi)(\tan\beta+\tan\phi)\tan^2\alpha+\tan\beta\tan\phi(\tan\delta+\tan\phi)\tan\alpha]$$

$$+\left[\frac{2V}{\gamma(h^2-d_c^2)}\right][1-\tan\delta\tan\phi-\tan\beta(\tan\delta+\tan\phi)][\tan^2\alpha-(\tan\beta+\tan\phi)\tan\alpha]$$

$$+\left[\frac{2V}{\gamma(h^2-d_c^2)}\right][1-\tan\delta\tan\phi-\tan\beta(\tan\delta+\tan\phi)]\tan\beta\tan\phi=0$$

$$n\frac{dm}{d\alpha}-m\frac{dn}{d\alpha}=\left[\frac{2V}{\gamma(h^2-d_c^2)}\right](1+\tan^2\phi)\tan^2\alpha-\left[\frac{4V}{\gamma(h^2-d_c^2)}\right]\tan\beta(1+\tan^2\phi)\tan\alpha$$

$$+\left[\frac{2V}{\gamma(h^2-d_c^2)}\right]\tan^2\beta(1+\tan^2\phi)=0$$

<u>The "c" term:</u>

$$m=-\frac{2c}{\gamma(h+d_c)}-\frac{2c\tan^2\alpha}{\gamma(h+d_c)},\quad \frac{dm}{d\alpha}=-\frac{4c\tan\alpha}{\gamma(h+d_c)}$$

$$\frac{dn}{d\alpha}=2(\tan\delta+\tan\phi)\tan\alpha+[1-\tan\delta\tan\phi-\tan\beta(\tan\delta+\tan\phi)]$$

$$n\frac{dm}{d\alpha}=-\left[\frac{4c}{\gamma(h+d_c)}\right](\tan\delta+\tan\phi)\tan^3\alpha$$

$$-\left[\frac{4c}{\gamma(h+d_c)}\right][1-\tan\delta\tan\phi-\tan\beta(\tan\delta+\tan\phi)]\tan^2\alpha$$

$$+\left[\frac{4c}{\gamma(h+d_c)}\right][\tan\beta(1-\tan\delta\tan\phi)]\tan\alpha$$

$$m\frac{dn}{d\alpha}=-\left[\frac{4c}{\gamma(h+d_c)}\right](\tan\delta+\tan\phi)\tan^3\alpha$$

$$-\left[\frac{2c}{\gamma(h+d_c)}\right][1-\tan\delta\tan\phi-\tan\beta(\tan\delta+\tan\phi)]\tan^2\alpha$$

$$-\left[\frac{4c}{\gamma(h+d_c)}\right](\tan\delta+\tan\phi)\tan\alpha$$

$$-\left[\frac{2c}{\gamma(h+d_c)}\right][1-\tan\delta\tan\phi-\tan\beta(\tan\delta+\tan\phi)]$$

$$n\frac{dm}{d\alpha} - m\frac{dn}{d\alpha} = -\left[\frac{2c}{(h+d_c)}\right][\,1 - \tan\delta\tan\phi - \tan\beta(\tan\delta + \tan\phi)]\tan^2\alpha$$

$$+\left[\frac{4c}{\gamma(h+d_c)}\right][\tan\beta(1 - \tan\delta\tan\phi) + \tan\delta + \tan\phi\,]\tan\alpha$$

$$+\left[\frac{2c}{\gamma(h+d_c)}\right][1 - \tan\delta\tan\phi - \tan\delta\tan\beta - \tan\beta\tan\phi\,] = 0$$

<u>Combining "W", "V", and "c" terms:</u>

$$-[\tan\phi + \tan\delta - \left[\frac{2V}{\gamma(h^2-d_c^2)}\right](1+\tan^2\phi) + \left[\frac{2c}{\gamma(h+d_c)}\right](1 - \tan\delta\tan\phi - \tan\beta[\tan\delta + \tan\phi])]\tan^2\alpha$$

$$+[\,2\tan\phi(\tan\delta + \tan\phi) - \left[\frac{4V}{\gamma(h^2-d_c^2)}\right]\tan\beta(\,1+\tan^2\phi\,)$$

$$+\left[\frac{4c}{\gamma(h+d_c)}\right](\tan\beta + \tan\phi + \tan\delta\,[\,1 - \tan\beta\tan\phi\,]\,)\,]\tan\alpha$$

$$+[\,\tan\phi - \tan\beta - (\,\tan\delta + \tan\beta\,)\tan^2\phi + \left[\frac{2V}{\gamma(h^2-d_c^2)}\right]\tan^2\beta(\,1+\tan^2\phi)$$

$$+\left[\frac{2c}{\gamma(h+d_c)}\right][\,1 - \tan\delta\tan\phi - \tan\beta(\,\tan\delta + \tan\phi\,)\,] = 0$$

In order to make the above equation less cumbersome, let:

$$1 - \tan\delta\tan\phi - \tan\beta(\,\tan\delta + \tan\phi\,) = r$$
$$\tan\beta + \tan\phi + \tan\delta(\,1 - \tan\beta\tan\phi\,) = s$$
$$\tan\phi - \tan\beta - (\,\tan\delta + \tan\beta\,)\tan^2\phi = t$$

Then the equation becomes:

$$-\left[\tan\phi + \tan\delta - \left(\frac{2V}{\gamma(h^2-d_c^2)}\right)(\,1+\tan^2\phi\,) + \left(\frac{2c}{\gamma(h+d_c)}\right)r\right]\tan^2\alpha$$

$$+\left[2\tan\phi(\tan\delta + \tan\phi) - \left(\frac{4V}{\gamma(h^2-d_c^2)}\right)\tan\beta(\,1+\tan^2\phi) + \left(\frac{4c}{\gamma(h+d_c)}\right)s\right]\tan\alpha$$

$$+\left[t + \left(\frac{2V}{\gamma(h^2-d_c^2)}\right)\tan^2\beta(\,1+\tan^2\phi) + \left(\frac{2c}{\gamma(h+d_c)}\right)r\right] = 0$$

Denoting the coefficient of $\tan^2\alpha$ as $-A$, the coefficient of $\tan\alpha$ as AC_1, and the constant as AC_2 we have the following quadratic equation:

$$-A\tan^2\alpha + AC_1\tan\alpha + AC_2 = 0$$

Dividing this by $-A$ gives:

$$\tan^2\alpha - C_1\tan\alpha - C_2 = 0$$

The solution for $\tan\alpha$ then is:

$$\tan\alpha = \frac{C_1 + \sqrt{C_1^2 + 4C_2}}{2}$$

$$\textit{alternately}$$

$$\alpha = \tan^{-1}\left(\frac{C_1 + \sqrt{C_1^2 + 4C_2}}{2}\right)$$

where

$$A = \tan\phi + \tan\delta - \left(\frac{2V}{\gamma(h^2 - d_c^2)}\right)(1 + \tan^2\phi) + \left(\frac{2c}{\gamma(h + d_c)}\right)r$$

$$C_1 = \frac{2\tan\phi(\tan\delta + \tan\phi) - \left(\frac{4V}{\gamma(h^2 - d_c^2)}\right)\tan\beta(1 + \tan^2\phi) + \left(\frac{4c}{\gamma(h + d_c)}\right)s}{A}$$

$$C_2 = \frac{t + \left(\frac{2V}{\gamma(h^2 - d_c^2)}\right)\tan^2\beta(1 + \tan^2\phi) + \left(\frac{2c}{\gamma(h + d_c)}\right)r}{A}$$

Appendix F
Effect of Vertical Shear on the Stability of Gravity Walls

F-1. Introduction

This appendix provides a simplified design procedure to calculate the vertical shear (downdrag) force on gravity walls due to backfill settlement. The downdrag force represents the total shear force on a vertical plane extending through the backfill from the heel of the wall. The downdrag force improves stability and therefore should be considered in those cases where the wall does not meet stability requirements if the downdrag is ignored. Downdrag may be mobilized either due to the movement of the wall as the backfill is placed, or due to settlement of the backfill; however, this appendix is limited to the case of non-moving walls with downdrag forces induced by backfill settlement. Backfill settlement introduces shear strains in the backfill material, which produce the vertical downdrag force. Once the horizontal force (F_H) has been determined and once the downdrag force (F_V) is computed using the procedures described in this appendix, the mobilized friction angle (δ) along a vertical plane extending through the backfill from the heel of the wall can be determined as $\delta = \tan^{-1} F_V/F_H$. It is generally not necessary to determine the value of δ; however, this may be useful if general wedge solutions are to be applied subsequent to determining the value of F_V.

F-2. Background

a. Previous design practice. In most cases of massive retaining walls constructed on rock foundations, movements of the wall and backfill are not sufficient to fully mobilize the shear resistance of the soil. Past practice has been to determine lateral earth pressures acting against the back of the gravity wall and set the interface friction between the wall and the backfill equal to zero. Zero interface friction along the back of the wall corresponds to a zero shear force along the back of the wall.

b. Finite element analysis. To develop an improved understanding of the interaction between gravity walls, their foundations, and their backfills, an investigation using finite-element analyses was conducted (Ebeling et al. 1990 and 1992; Ebeling, Duncan, and Clough 1990). The analyses demonstrated that the backfill settles relative to the wall and develops downward-shear loads on the wall. The magnitude of the vertical shear force (F_V) is expressed in terms of a vertical shear coefficient (K). This coefficient is related to the shear force on the vertical plane through the heel of a wall by the following equation (note that some of the nomenclature used in this appendix has been altered from the nomenclature used in referenced source material):

$$F_V = K \left(\frac{\gamma H^2}{2} \right) \qquad \text{(F-1)}$$

where: γ = unit weight of backfill
H = depth of backfill against wall

Figure F-1 shows the results of finite-element analyses of four walls founded on rock and retaining a moist backfill, where values of K range from 0.09 to 0.21. The analyses indicated that rock founded gravity walls would move only slightly during the placement of the toe fills and backfills. As a result, the earth pressures on the backs and fronts of the walls are close to those that exist at rest. Even so, settlement of the backfill as it is placed behind the wall is sufficient to generate a significant amount of shear force. Parametric studies demonstrated that the most important factors influencing the value of K for concrete gravity walls on rock foundations are the depth of the backfill, the stiffness of the backfill, the inclination of the back of the wall, and the number of steps in the back of the wall. The following trends were observed:

- For low walls, the value of K increases with increasing wall height because more backfill compression occurs due to self-weight of the backfill. The resulting increase in differential movement between backfill and the wall causes a greater portion of the interface strength to be mobilized. This process approaches a limiting condition for high walls as the interface strength becomes fully mobilized over most of the wall-backfill contact area.

- As the stiffness of the backfill increases, backfill compression decreases, and the wall height necessary to mobilize the full interface strength increases. For low walls with vertical back sides, the value of K decreases as the backfill stiffness increases.

- The value of K decreases as the back side of the wall becomes inclined away from the backfill.

- The value of K is greater for a wall with a stepped-back side than for a wall with a smooth-back side at the same average slope.

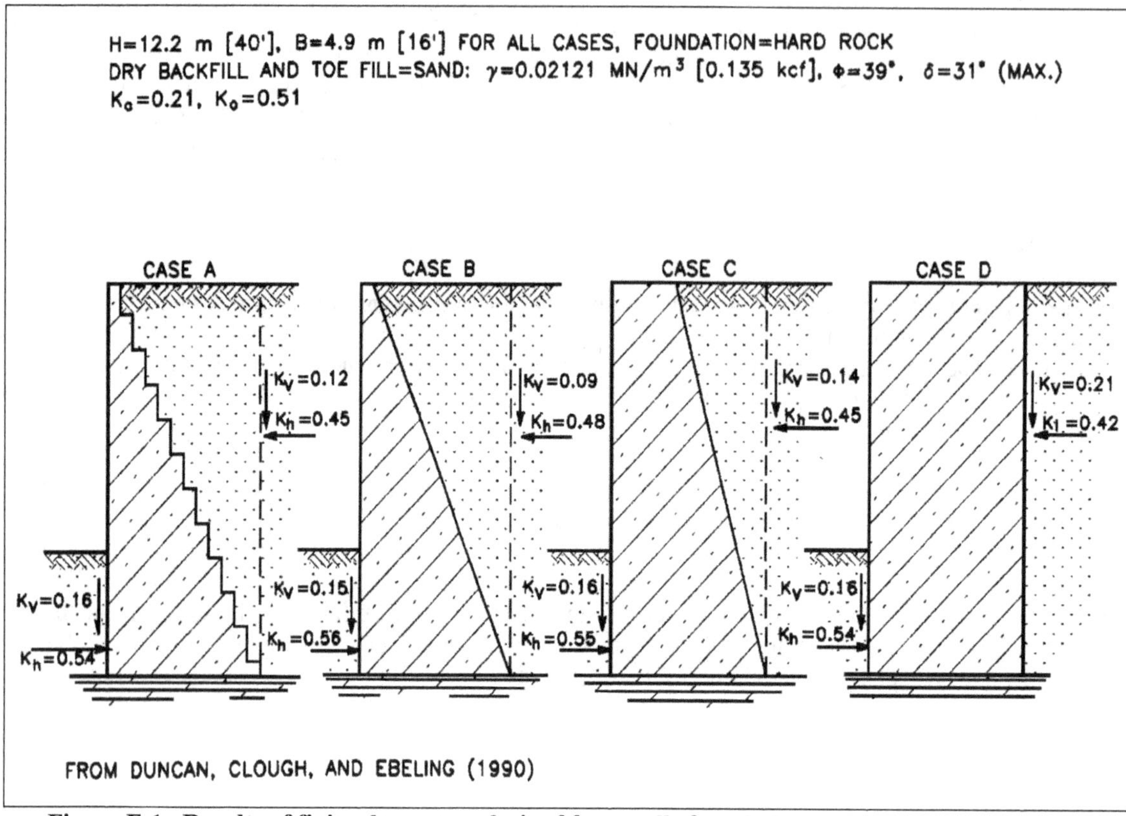

Figure F-1. Results of finite element analysis of four walls founded on rock-retaining dry backfill (K = value of K_V shown in this figure)

c. Field and model wall measurements. Shear loads have been reported for several instrumented walls (Duncan, Clough, and Ebeling 1990 and Filz, Duncan, and Ebeling 1997), including a lock wall 30.2 m (99.1 ft) in height and founded on rock (Hilmer 1986). Measurements at the lock wall are reported over a 6-year period. Mobilized interface friction at the lock wall fluctuates seasonally and with changes in the water level inside the lock. However, the data indicate that the shear force is persistent over the 6-year period and does not decay with time. According to a conservative interpretation of the data, the minimum value of K during the 6-year period is about 0.18. In a research program conducted at Virginia Polytechnic Institute and State University (Filz and Duncan 1992), both the horizontal earth force and the vertical shear force along the vertical back side of a 2.1 m (7 ft) high rigid retaining wall were measured. The research program included 16 tests using compacted fine sand (Unified Soil Classification SP) and compacted non-plastic silty sand (SM) as backfill. Measured values of K ranged from 0.11 to 0.23 (Table 8.9 in Filz and Duncan 1992 or Table 6.2 in Filz, Duncan, and Ebeling 1997). The more compressible backfills exhibited higher K values. The compacted backfills were left in place for periods ranging from 1 to 14 days after completion of backfilling. Values of K tended to increase with time.

d. Benefits. Filz, Duncan, and Ebeling (1997) present an example calculation using vertical shear for a 30-ft-high, step-tapered, rock-founded, gravity wall retaining dense sand with surcharge (no ground water table). This example compares the result with a conventional design and shows a 14 percent reduction in base width by including vertical shear, without compromising the design safety requirements.

e. Limits on use of vertical shear. The vertical shear force can be incorporated in conventional equilibrium calculations, subject to the following limitations:

- When a toe fill of significant height exists, a vertical shear force at the toe should be included in the equilibrium calculations if a vertical shear force was applied to the back of the wall. Neglecting the shear force at the toe could result in unconservative estimates of the base contact area and the maximum bearing pressure on the foundation.

- The vertical displacements within the foundation during construction of the wall and placement of the backfill are negligible when compared with the vertical settlement within the backfill due to self-weight. Gravity walls founded on competent rock foundations satisfy this criterion.

- The backfill does not creep. Compacted soils classified as SW, SP, GW, and GP according to the Unified Soil Classification System (ASTM 1990) do not experience significant creep movements, nor do select SM backfills with non-plastic fines that do not creep.

- No special features that reduce or eliminate interface friction exist along the interface between the back of the wall and the backfill. Examples of special features that would reduce interface friction include bituminous coatings and synthetic barriers with low interface friction values.

- The interface between the back side of the wall and the backfill is capable of developing friction values of $\delta > 0.7\ \varphi$, where φ is the effective angle of internal friction for the backfill. This is satisfied by SW, SP, GW, and GP soils compacted against concrete walls. It is also satisfied by SM soils with non-plastic fines compacted against concrete walls.

- The water pressures are hydrostatic within the backfill, and the rise in water table is concurrent with placement of the soil lifts.

F-3. Procedures for Calculating the Vertical Shear Force

Two basic procedures for computing the magnitudes of shear loads along the backs of gravity walls are described in this section: a simplified procedure and a complete soil-structure interaction analysis using finite elements. These procedures are intended only as guidelines and are not intended to replace judgment by the project engineers.

a. Simplified procedure for level backfill, with no surcharge. For level backfill, with no permanent surcharge, the downdrag on a vertical plane through the heel of the wall can be computed as:

$$F_v = K\left[\frac{1}{2}\gamma_{moist}(D_1)^2 + \gamma_{moist}(D_1 D_2) + \frac{1}{2}\gamma_b(D_2)^2\right] \qquad \text{(F-2)}$$

where: K = the vertical earth force coefficient
D_1 = thickness of backfill above the hydrostatic water table
D_2 = thickness of submerged backfill above the base of the wall
$H = D_1 + D_2$ = total height of backfill
γ_{moist} = moist unit weight of backfill (above the water table)
γ_{sat} = saturated unit weight of submerged backfill
γ_w = unit weight of water
γ_b = buoyant unit weight of submerged backfill = $\gamma_{sat} - \gamma_w$

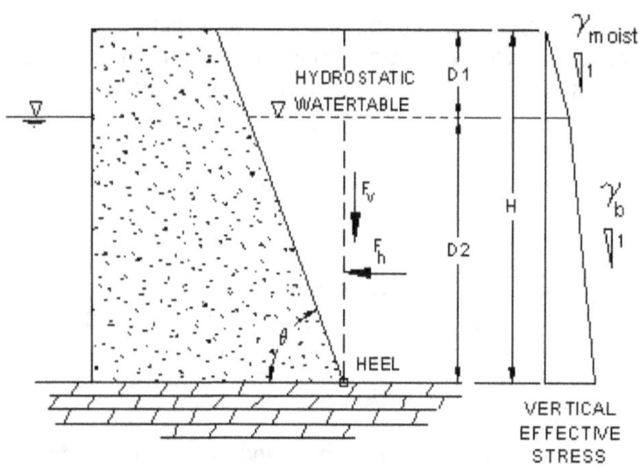

Figure F-2 Forces acting on vertical plane at heel of wall

This equation assumes that the water pressures are hydrostatic within the backfill and the rise in water table is concurrent with the placement of soil lifts. The term in the bracket represents the integral of the vertical effective overburden stress shown in Figure F-2. (The case of walls in which there is a post-construction rise in groundwater level in the backfill will be discussed later in this appendix). For the case of dry backfill, D_2 equals zero and Equation F-2 reduces to Equation F-1. In Equation F-2, the value of K is obtained from Equation F-3:

$$K = (1 - C_\theta C_N) K_V \qquad \text{(F-3)}$$

where: K_V = from Figure F-3,
C_N = correction factor for the number of steps in the backfill side, from Figure F-4,
C_θ = correction factor for inclination of the backfill side, from Figure F-4,

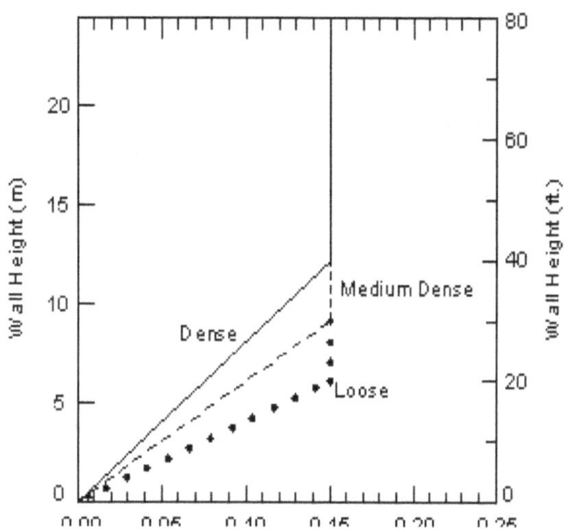

Figure F-3 Design values of K_v for gravity walls founded on rock, with a vertical face against the backfill

Figure F-3 shows that the value of K_V increases with increasing wall height until a limiting value of 0.15 is reached, and that the limiting value develops at lower heights for walls with loose backfill than for walls with dense backfill. The recommended design value is well below the actual value indicated by measurements and analyses. It was selected to be conservative because it is a simplified design procedure. Even with this conservative selection of the design value, significant economies can be obtained by including the vertical shear force in design.

b. Simplified procedure for surcharges. In the case of rock-founded gravity walls with surcharges on the backfill, F_v is calculated using the following procedure. This procedure can also be used to simulate the effects of sloping backfill surfaces, where the additional weight of backfill above some horizontal plane can be represented as an idealized surcharge. Terminology is the same as used for flat backfills without surcharges, unless otherwise noted.

$$F_v = K\,[\,\tfrac{1}{2}\,\gamma_{moist}\,(D_1)^2 + \gamma_{moist}(D_1\,D_2) + \tfrac{1}{2}\,\gamma_b\,(D_2)^2] + K'\;C_s\,(q)\,H \qquad \text{(F-4)}$$

where all variables are identical to equation F-2, except:

$K' = 1.4$ K, but $K' \leq 0.15$

C_S = the correction factor for an inclined backfill surface, from Figure F-3,

q = surcharge pressure on the backfill (for sloping backfills $q = \gamma_{moist}\ \Delta H$).

The reason for use of K' in the last term of this equation, rather than K, is that the surcharge load increases the backfill settlement, thus resulting in greater vertical shears. This procedure should only be used for permanent surcharges; temporary surcharges might not be sufficient to significantly increase backfill settlement.

c. Post-construction rise in ground water. Equations F-2 and F-4 assume that the rise in water table within the backfill is concurrent with placement of the soil lifts. In many cases, the backfill is placed before submergence takes place. For walls in which the construction of the wall and the placement of "dry" (or more precisely, moist) backfill precedes a rise in the ground water level, a soil-structure interaction analysis procedure is generally used to compute the values of F_v for both a dry backfill condition and a post-construction rise in ground water condition. In these situations, a rebound of the soil can occur. This results in a reduction in the effective stress and can result in a reduction in the shear force Fv. This occurred in analysis of the soil-founded U-frame lock at Red River Lock No. 1 (Ebeling et al. 1993, Ebeling and Mosher 1996). The soil-structure-foundation interaction analysis of the new roller-compacted concrete (RCC) lock (rock founded) at McAlpine Locks by Ebeling and Wahl (1997) is another example of this type of analysis. Both structures were constructed in the dry, followed by partial submergence of the lock sites. However, a simplified procedure is also available to compute values for F_v after backfilling and after a post-construction rise in ground water level. This procedure is based on data from the soil-structure-foundation interaction analysis of the new roller-compacted lock wall at McAlpine Locks (Ebeling and Wahl 1997). Due to limited data, this procedure is restricted to level backfills with no surcharge. Two values for F_v are computed for the two stages of wall construction. The first stage being that which occurs after backfilling is complete. The second stage being that which occurs after a post-construction rise in the ground water level.

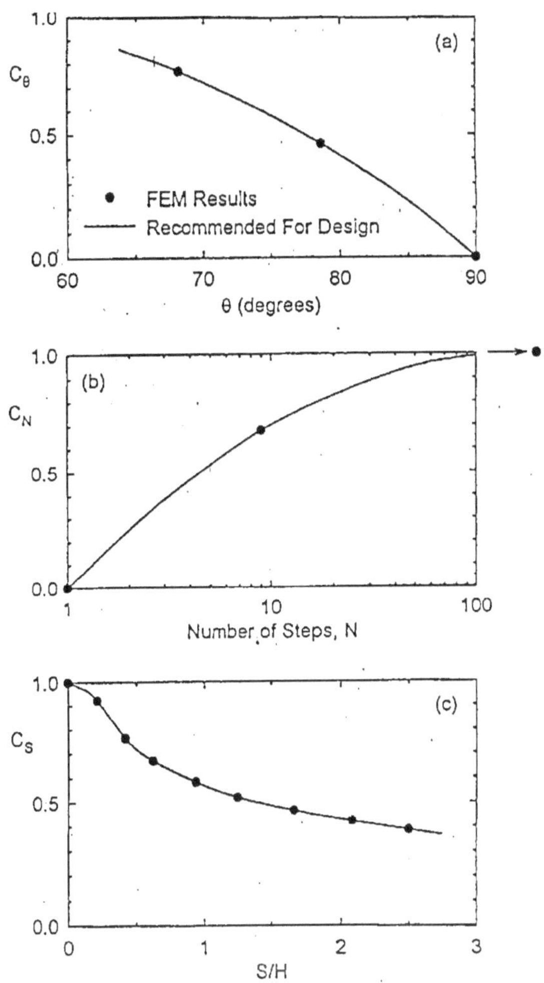

Figure F-4 Correction factors C_θ, C_N, and C_S (geometry definitions are illustrated in Figure F-5)

(1) Stage 1. F_v after backfilling is computed using Equation F-1, or Equation F-2 with D_2 set equal to zero.

(2) Stage 2. After the post-construction rise in the ground water level, F_v is calculated using equation F-2, but substituting $K\,C_{wt}$ in place of K. The value for the correction factor C_{wt} is given by:

$$C_{wt} = \left\{1 - \frac{D_2}{H}\right\} \tag{F-5}$$

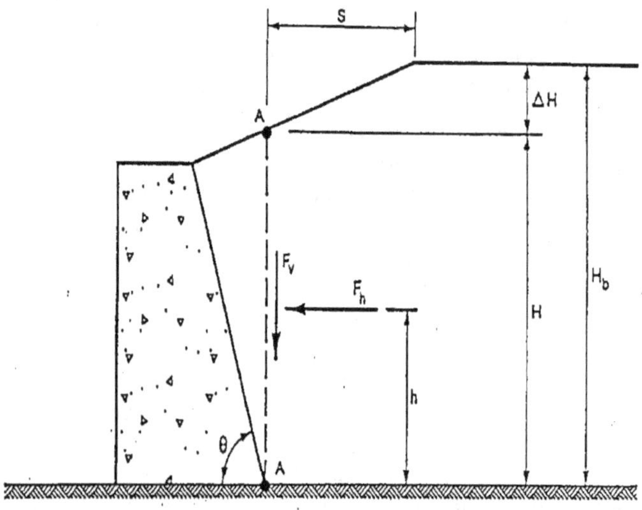

(a) Mass concrete wall with a planar, sloping backside and an inclined backfill surface

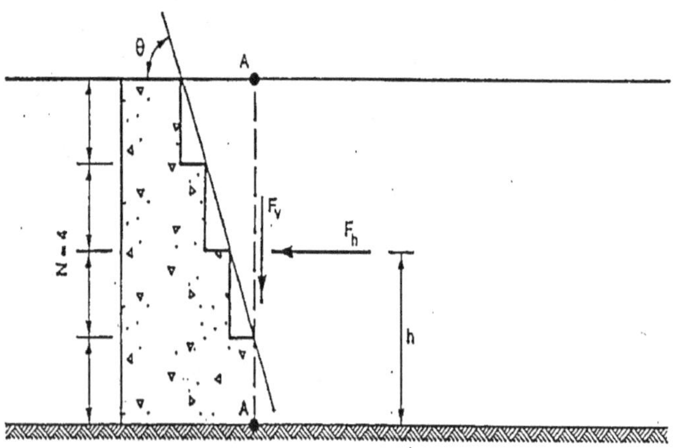

(b) Mass concrete wall with a stepped backside

Figure F-5 Geometry definition

d. Soil-structure interaction analysis. A complete soil-structure interaction (SSI) analysis is recommended for those structures for which the simplified procedure is not applicable, or for those cases in which a more precise evaluation of the shear force is required. Soil-structure analyses are recommended for U-frame locks, retaining structures founded on soils, and structures with complicated geometry. This can be accomplished using a finite element program such as SOILSTRUCT (Ebeling, Peters, and Clough 1992). Unlike conventional equilibrium procedures, an SSI analysis does not require the use of predetermined pressure distributions between the soil and the wall. Instead, it allows for development of these pressures through soil-structure interaction by simulating the staged construction that occurs. The computer program SOILSTRUCT can model the nonlinear stress-strain behavior of the soil and allow for relative movement between the soil and the structure by incorporating interface elements in the mesh. SSI analyses are also especially useful for analyzing retaining structures founded on either soils or compressible rock foundations. Differential settlements within the foundation affect the magnitude of the shear force that the backfill exerts on the wall. The SSI analysis procedure has been successfully used for a wide variety of problems, including the Port Allen and Old River locks (Clough and Duncan 1969) and, more recently, the lock at Red River Lock and Dam No. 1 (Ebeling et al. 1993, Ebeling and Mosher 1996, and Ebeling, Peters, and Mosher 1997). Additional information on soil-structure Interaction analysis can be found in Ebeling, Pace, Morrison (1997).

Appendix G
Earthquake Forces from Backfill

G-1. General

For hydraulic structures, which are able to yield laterally during an earthquake, the calculation of increased earth pressures induced by earthquakes can be approximated by the approaches outlined below. In addition, the inertial forces of the structure, plus that portion of the adjacent earth and/or water, which is assumed to act with the structure, should be included.

G-2. Mononobe-Okabe Analysis

This analysis is an extension of the Coulomb sliding-wedge theory taking into account horizontal and vertical inertial forces acting on the soil. The analysis is described in detail by Seed and Whitman (1970) and Whitman and Liao (1985).

a. Assumptions. The following assumptions are made by the Mononobe-Okabe analysis:

- The structure is free to yield sufficiently to enable full soil strength or active pressure conditions to be mobilized.
- The backfill is completely above or completely below the water table, unless the top surface is horizontal, in which case the backfill can be partially saturated.
- The backfill is cohesionless.
- The top surface is planar (not irregular or broken).
- Any surcharge is uniform and covers the entire surface of the soil wedge.
- Liquefaction is not a problem.

b. Equations. Equilibrium considerations of the soil wedge on the driving and resisting sides lead to the following Mononobe-Okabe equations for computing the active and passive forces exerted by the soil on the structure when the soil mass is at the point of failure (total shear resistance mobilized) along the slip plane of the Mononobe-Okabe wedge shown in Figure G-1:

For driving (active) wedges (Figure G-1a),

$$P_{AE} = \frac{1}{2} K_{AE} \gamma (1 - k_v) h^2 \qquad \text{(G-1)}$$

$$K_{AE} = \frac{\cos^2(\phi - \psi - \theta)}{\cos\psi \cos^2\theta \cos(\psi + \theta + \delta)\left[1 + \sqrt{\dfrac{\sin(\phi + \delta)\sin(\phi - \psi - \beta)}{\cos(\beta - \theta)\cos(\psi + \theta + \delta)}}\right]^2} \;\cdots\cdots \qquad \text{(G-2)}$$

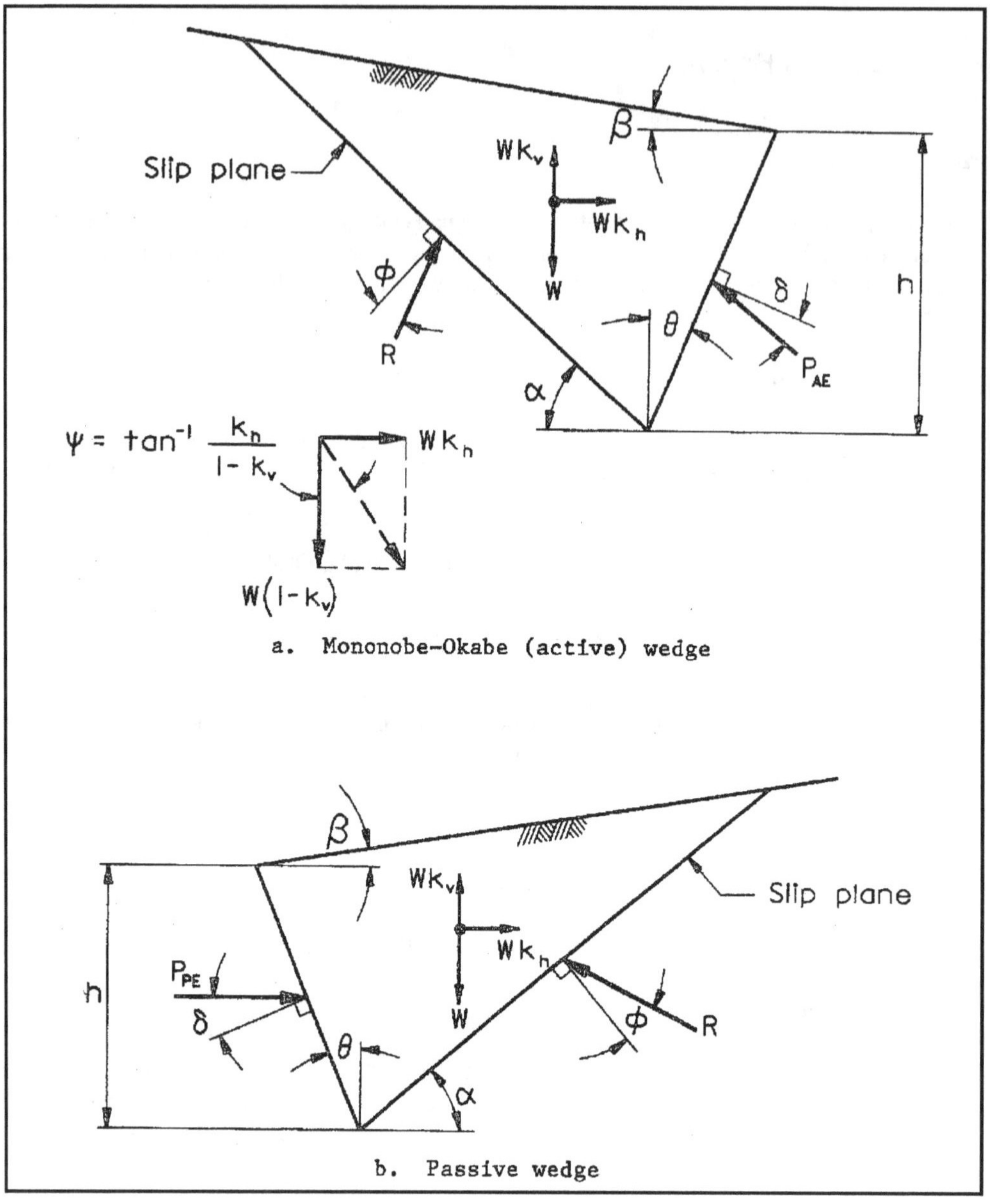

a. Mononobe-Okabe (active) wedge

b. Passive wedge

Figure G-1. Driving and resisting seismic wedges, no saturation

For resisting (passive) wedges (Figure G-1b),

$$P_{PE} = \frac{1}{2} K_{PE} \gamma (1 - k_v) h^2 \tag{G-3}$$

$$K_{PE} = \frac{\cos^2 (\phi - \psi - \theta)}{\cos \psi \cos^2 \theta \cos (\psi - \theta + \delta) \left[1 - \sqrt{\frac{\sin (\phi + \delta) \sin (\phi - \psi + \beta)}{\cos (\beta - \theta) \cos (\psi - \theta + \delta)}} \right]^2} \tag{G-4}$$

P_{AE} and P_{PE} are the combined static and dynamic forces due to the driving and resisting wedges, respectively. The equations are subject to the same limitations that are applicable to Coulomb's equations. Definitions of terms are as follows:

γ = unit weight of soil
k_v = vertical acceleration in g's
h = height of structure
ϕ = internal friction angle of soil
$\psi = \tan\left(\frac{k_h}{1-k_v}\right)$ = seismic inertia angle
k_h = horizontal acceleration in g's
θ = inclination of interface with respect to vertical (this definition of θ is different from θ 1in Coulomb's equations)
δ = soil-structure friction angle
β = inclination of soil surface (upward slopes away from the structure are positive)

(c) *Simplifying Conditions.* For the usual case where k_v, δ, and θ are taken to be zero, the equations reduce to:

$$K_{AE} = \frac{\cos^2(\phi - \psi)}{\cos^2\psi\left[1 + \sqrt{\frac{\sin\phi\,\sin(\phi - \psi - \beta)}{\cos\beta\,\cos\psi}}\right]^2} \quad \text{(G-5)}$$

$$K_{PE} = \frac{\cos^2(\phi - \psi)}{\cos^2\psi\left[1 - \sqrt{\frac{\sin\phi\,\sin(\phi - \psi + \beta)}{\cos\beta\,\cos\psi}}\right]^2} \quad \text{(G-6)}$$

where

$\psi = \tan^{-1}(k_h)$

and

$P_{AE} = ½\, K_{AE}\, \gamma h^2$
$P_{PE} = ½\, K_{PE}\, \gamma h^2$

For the case when the water table is above the backfill, P_{AE} and P_{PE} must be divided into static and dynamic components for computing the lateral forces. Buoyant soil weight is used for computing the static component below the water table, with the hydrostatic force added, and saturated soil weight is used for computing the dynamic component.

(d) *Observations.* General observations from using Mononobe-Okabe analysis are as follows:

(1) As the seismic inertia angle ψ increases, the values of K_{AE} and K_{PE} approach each other and, for a vertical backfill face ($\theta = 0$), become equal when $\psi = \phi$.

(2) The locations of P_{AE} and P_{PE} are not given by the Mononobe-Okabe analysis. Seed and Whitman (1970) suggest that the dynamic component ΔP_{AE} be placed at the upper one-third point, ΔP_{AE} being the difference between P_{AE} and the total active force from Coulomb's active wedge without the earthquake. The general wedge earthquake analysis described in paragraph 3-26c places the dynamic component ΔP_{AE} at the upper one-third point also, but

computes ΔP_{AE} as being the difference between P_{AE} and the total active force from the Mononobe-Okabe wedge. The latter method for computing ΔP_{AE}, which uses the same wedge for computing the static and dynamic components of P_{AE}, is preferred.

(3) The radical in the Mononobe-Okabe equation must be positive for a real solution to be possible, and for this it is necessary that $\phi \geq \psi + \beta$ for the driving wedges and $\phi \geq \psi - \beta$ for the resisting wedges. This is a limit to the horizontal acceleration coefficient for a soil wedge. The limiting condition for the driving wedge is:

$$k_h \leq (1 - k_v) \tan (\phi - \beta) \qquad \text{(G-7)}$$

and for the resisting wedge:

$$k_h \leq (1 - k_v) \tan (\phi + \beta) \qquad \text{(G-8)}$$

(4) Figure G-2a (Applied Technology Council 1981) shows the effect on the magnification factor F_T (equal to K_{AE}/K_A) on changes in the vertical acceleration coefficient k_v. Positive values of k_v have a significant effect for values of k_v greater than 0.2. The effect is greater than 10 percent above and to the right of the dashed line. For values of k_h of 0.2 or less, k_v can be neglected for all practical purposes.

(5) K_{AE} and F_T are also sensitive to variations in backfill slope, particularly for higher values of horizontal acceleration. This effect is shown in Figure G-2b.

G-3. General Wedge Earthquake Analysis

For many projects, all the Coulomb wedge assumptions are met, and the following wedge analysis should be used. The equations for the dynamic force given below for various conditions are simply the horizontal acceleration coefficient multiplied by the weight of the wedge defined by the critical slip-plane angle. See the example later in this chapter for more information.

a. Assumptions. The equations for determining the critical slip-plane angle for driving and resisting wedges subjected to a horizontal acceleration are developed with the following assumptions:

(1) The shear on the vertical face of the wedge is zero.

(2) The shear strength along the potential slip planes in the soil has not been mobilized to any extent, i.e., for static loading prior to an earthquake.

b. Equations for Cohesionless, Dry Backfill Above the Water Table. Driving and resisting forces for cohesionless, dry, sloping planar-surfaced backfill below the water table where k_v, δ, and $\theta = 0$ can be computed as follows:

(1) Static Components. The static components for a driving and resisting wedge are:

$$P_A = \frac{1}{2} K_A \gamma h^2 \qquad \text{(G-9)}$$

$$P_P = \frac{1}{2} K_P \gamma h^2 \qquad \text{(G-10)}$$

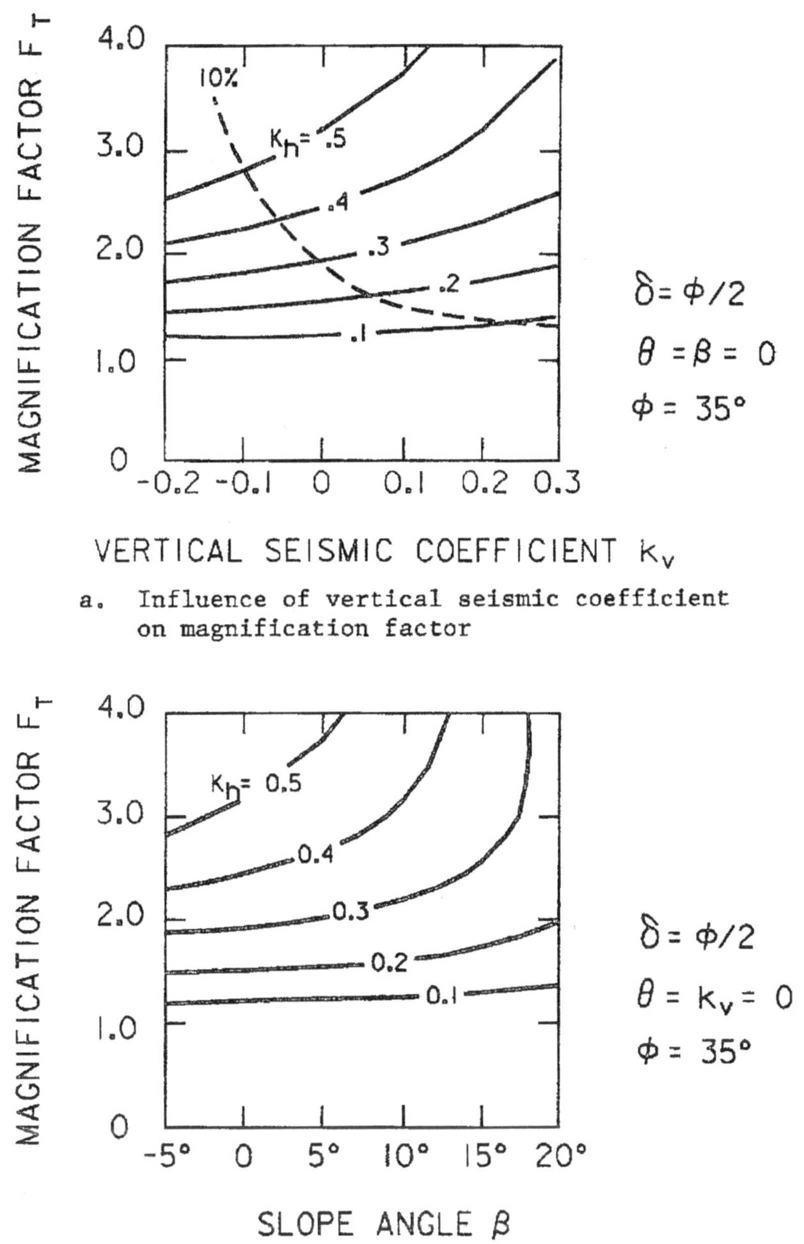

a. Influence of vertical seismic coefficient on magnification factor

b. Influence of backfill slope angle on magnification factor

Figure G-2. Influence of k_v and β on magnification factor (after Applied Technology Council 1981)

where

$$K_A = \left(\frac{1 - \tan\phi \cot\alpha}{1 + \tan\phi \tan\alpha}\right)\left(\frac{\tan\alpha}{\tan\alpha - \tan\beta}\right) \tag{G-11}$$

$$K_P = \left(\frac{1 + \tan\phi \cot\alpha}{1 - \tan\phi \tan\alpha}\right)\left(\frac{\tan\alpha}{\tan\alpha - \tan\beta}\right) \tag{G-12}$$

For an active wedge:

$$\alpha = \tan^{-1}\left(\frac{c_1 + \sqrt{c_1^2 + 4c_2}}{2}\right) \qquad \text{(G-13)}$$

$$c_1 = \frac{2\,(\tan\phi - k_h)}{1 + k_h\,\tan\phi} \qquad \text{(G-14)}$$

$$c_2 = \frac{\tan\phi\,(1 - \tan\phi\,\tan\beta) - (\tan\beta + k_h)}{\tan\phi\,(1 + k_h\,\tan\phi)} \qquad \text{(G-15)}$$

For a passive wedge:

$$\alpha = \tan^{-1}\left(\frac{-c_1 + \sqrt{c_1^2 + 4c_2}}{2}\right) \qquad \text{(G-16)}$$

$$c_1 = \frac{2\,(\tan\phi - k_h)}{1 + k_h\,\tan\phi} \qquad \text{(G-17)}$$

$$c_2 = \frac{\tan\phi\,(1 - \tan\phi\,\tan\beta) - (\tan\beta + k_h)}{\tan\phi\,(\,+ k_h\,\tan\phi)} \qquad \text{(G-18)}$$

If $k_v > 0$, replace γ with $(1 - k_v)\,\gamma$ and replace k_h with $\tan\phi$ or with $k_h/(1 - k_v)$.

(2) Dynamic Components. The dynamic component for each wedge is:

$$\Delta P_{AE} = \Delta P_{PE} = k_h\left[\frac{\gamma h^2}{2\,(\tan\alpha - \tan\beta)}\right] \qquad \text{(G-19)}$$

(3) Total Driving Force. The total driving force is:

$$P_{AE} = P_A + \Delta P_{AE} \qquad \text{(G-20)}$$

which, from the Mononobe-Okabe analysis, is equal to:

$$P_{AE} = \frac{1}{2}K_{AE}\,\gamma h^2 \qquad \text{(G-21)}$$

The line of action for P_{AE} may be found as:

$$Y_{AE} = \frac{P_A \left(\frac{h}{3}\right) + \Delta P_{AE} \left(\frac{2h}{3}\right)}{P_{AE}} \qquad \text{(G-22)}$$

It should be noted that for large values of k_h, which cause α to be small, P_A can be negative causing the line of action of P_{AE} to lie above the upper third point.

(4) Total Resisting Force. The total resisting force is:

$$P_{PE} = P_P - \Delta P_{PE} \qquad \text{(G-23)}$$

which, from the Mononobe-Okabe analysis, is equal to:

$$P_{PE} = \frac{1}{2} K_{PE}\, \gamma h^2 \qquad \text{(G-24)}$$

The line of action for P_{PE} may be found as:

$$Y_{PE} = \frac{P_P \left(\frac{h}{3}\right) - \Delta P_{PE} \left(\frac{2h}{3}\right)}{P_{PE}} \qquad \text{(G-25)}$$

c. Equations for Cohesionless Backfill with Water Table. Driving and resisting forces for cohesionless, sloping, planar-surfaced backfill with water table where k_v, δ and $\theta = 0$ can be computed as follows:

(1) Driving Force. The static components for a driving wedge are (see Figures G-3a and G-4a):

$$P_A = P_{A1} + P_{A2} = \frac{1}{2} K_A \gamma (h - h_s)^2 + \frac{1}{2} h_s \left[2K_A \gamma (h - h_s) + K_b \gamma_b h_s\right] \qquad \text{(G-26)}$$

$$P_{ws} = \frac{1}{2} \gamma_w h_s^2 \qquad \text{(G-27)}$$

and the dynamic components are (see Figures G-3a and G-4a):

$$\Delta P_{AE} = \Delta P_{AE1} + \Delta P_{AE2} = k_h \left[\frac{\gamma h^2}{2\,(\tan\alpha - \tan\beta)}\right] + k_h \left[\frac{(\gamma_s - \gamma) h_s^2}{2 \tan\alpha}\right] \qquad \text{(G-28)}$$

giving a total force of:

$$P_{AE} = P_A + P_{ws} + \Delta P_{AE} \qquad \text{(G-29)}$$

where

γ_s = saturated unit weight of fill
γ = moist unit weight of fill
γ_b = buoyant unit weight of fill
γ_w = unit weight of water

$$K_A = \left(\frac{1 - \tan\phi \cot\alpha}{1 + \tan\phi \tan\alpha}\right)\left(\frac{\tan\alpha}{\tan\alpha - \tan\beta}\right)$$

$$K_b = \left(\frac{1 - \tan\phi \cot\alpha}{1 + \tan\phi \tan\alpha}\right)\left[1 + \left(\frac{\tan\alpha}{\tan\alpha - \tan\beta} - 1\right)\frac{\gamma}{\gamma_b}\right]$$

and α is defined in Equation G-13.

(2) Resisting Force. The static components for the resisting wedge are (see Figures G-3b and G-4b):

$$P_P = P_{P1} + P_{P2} = \frac{1}{2} K_P \gamma (h - h_s)^2 + \frac{1}{2} h_s \left[2K_P \gamma (h - h_s) + K_b \gamma_b h_s\right] \quad \text{(G-30)}$$

$$P_{ws} = \frac{1}{2} \gamma_w h_s^2 \quad \text{(G-31)}$$

and the dynamic components are:

$$\Delta P_{PE} = \Delta P_{PE1} + \Delta P_{PE2} = k_h \left[\frac{\gamma h^2}{2\,(\tan\alpha - \tan\beta)}\right] + k_h \left[\frac{(\gamma_s - \gamma) h_s^2}{2 \tan\alpha}\right] \quad \text{(G-32)}$$

giving a total force of:

$$P_{PE} = P_P + P_{ws} - \Delta P_{PE} \quad \text{(G-33)}$$

where γ, γ_b, γ_s, and γ_w are defined in paragraph 3-26c(3)(a),

and

$$K_P = \left(\frac{1 + \tan\phi \cot\alpha}{1 - \tan\phi \tan\alpha}\right)\left(\frac{\tan\alpha}{\tan\alpha - \tan\beta}\right) \quad \text{(G-34)}$$

$$K_b = \left(\frac{1 + \tan\phi \cot\alpha}{1 - \tan\phi \tan\alpha}\right)\left[1 + \frac{\tan\alpha}{\tan\alpha - \tan\beta} - 1\frac{\gamma}{\gamma_b}\right] \quad \text{(G-35)}$$

and the equations for α, are given in Equation G-16.

d. Equations for Cohesive Backfill with Water Table. Driving and resisting forces for a cohesive, sloping, planar-surfaced backfill with water table where k_v, δ, and $\theta = 0$ can be computed as follows:

(1) Driving Force. The static components for the driving wedge are (see Figure G-5a):

$$P_A = P_{A1} + P_{A2} = \frac{1}{2} K_A \gamma \left[(h - d_c) - h_s \right]^2 + \frac{1}{2} h_s \left[2K_A \gamma (h - d_c - h_s) + K_b \gamma_b h_s \right] \tag{G-36}$$

$$P_{ws} = \frac{1}{2} \gamma_w h_s^2 \tag{G-37}$$

and the dynamic components are (see Figure G-5a):

$$\Delta P_{AE} = \Delta P_{AE1} + \Delta P_{AE2} = k_h \left[\frac{\gamma (h^2 - d_c^2)}{2 \, (\tan \alpha - \tan \beta)} \right] + k_h \left[\frac{(\gamma_s - \gamma)^2 \, h_s^2}{2 \tan \alpha} \right] \tag{G-38}$$

giving a total force of:

$$P_{AE} = P_A + P_{ws} + \Delta P_{AE} \tag{G-39}$$

where

γ = moist unit weight of fill
γ_b = buoyant unit weight of fill
γ_s = saturated unit weight of fill
γ_w = unit weight of water

$$K_A = \left(\frac{1 - \tan \phi \cot \alpha}{1 + \tan \phi \tan \alpha} \right) \left(\frac{\tan \alpha}{\tan \alpha - \tan \beta} \right) \tag{G-40}$$

$$K_b = \left(\frac{1 - \tan \phi \cot \alpha}{1 + \tan \phi \tan \alpha} \right) \left[1 + \left(\frac{\tan \alpha}{\tan \alpha - \tan \beta} - 1 \right) \frac{\gamma}{\gamma_b} \right] \tag{G-41}$$

$$\alpha = \tan^{-1} \left(\frac{c_1 + \sqrt{c_1^2 + 4c_2}}{2} \right) \tag{G-42}$$

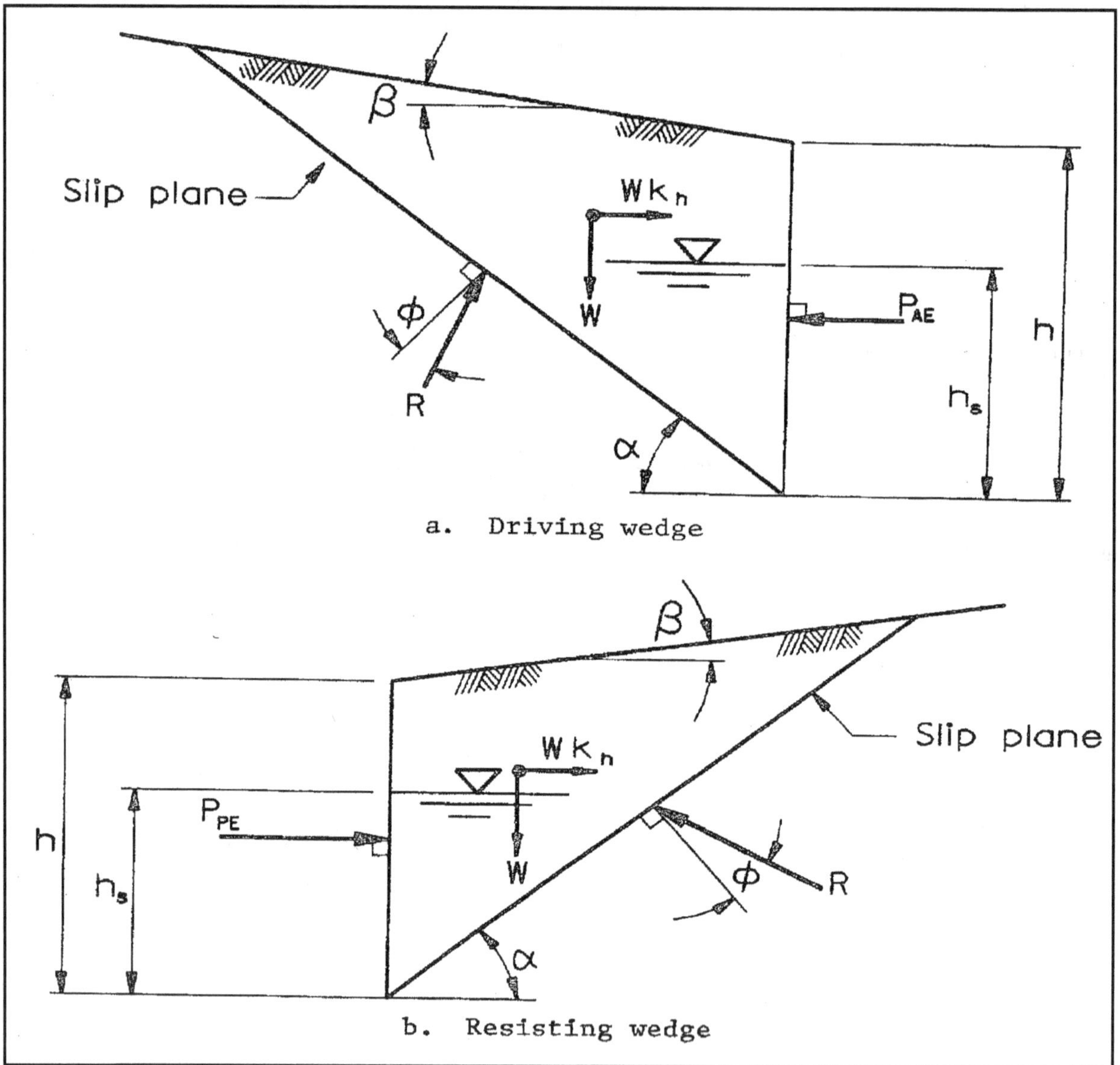

Figure G-3. Seismic wedges, water table within wedge

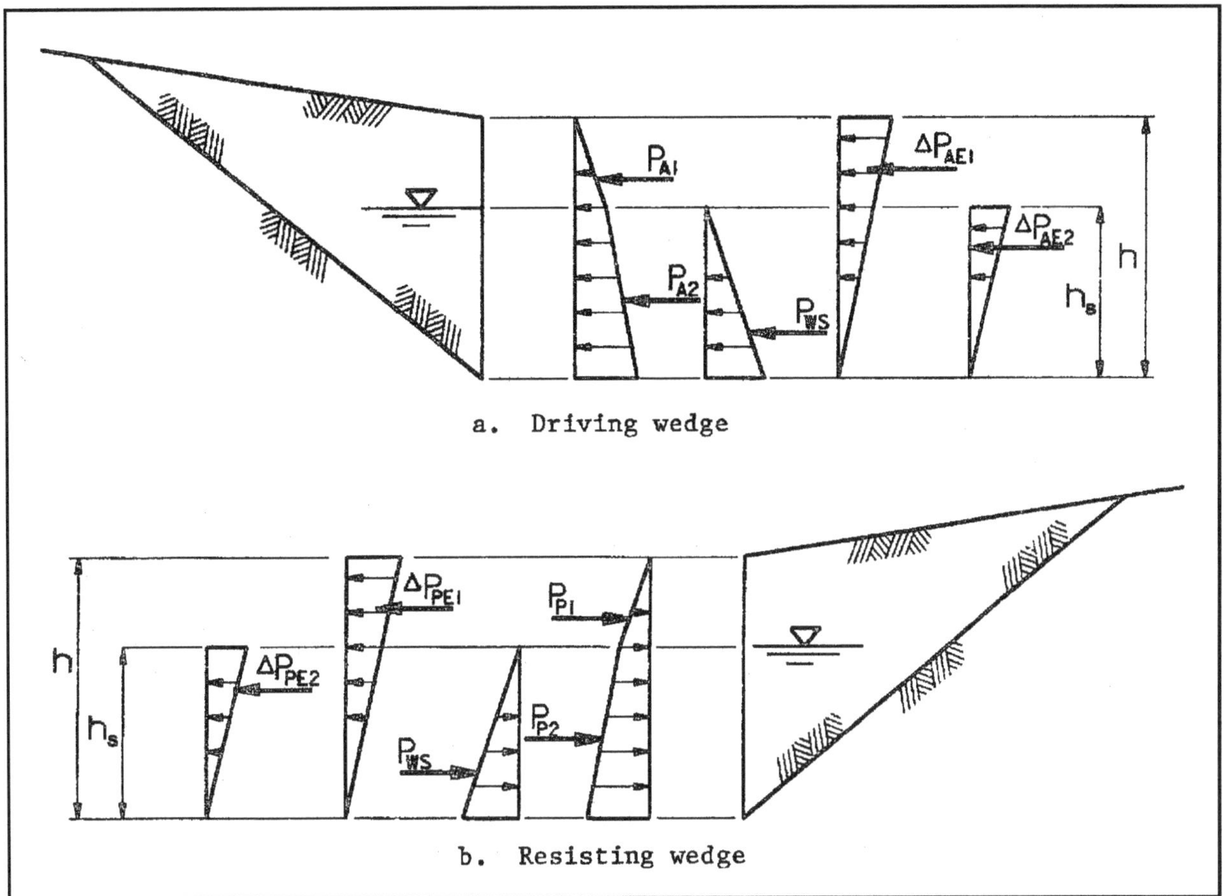

Figure G-4. Static and dynamic pressure diagrams, water table within wedge

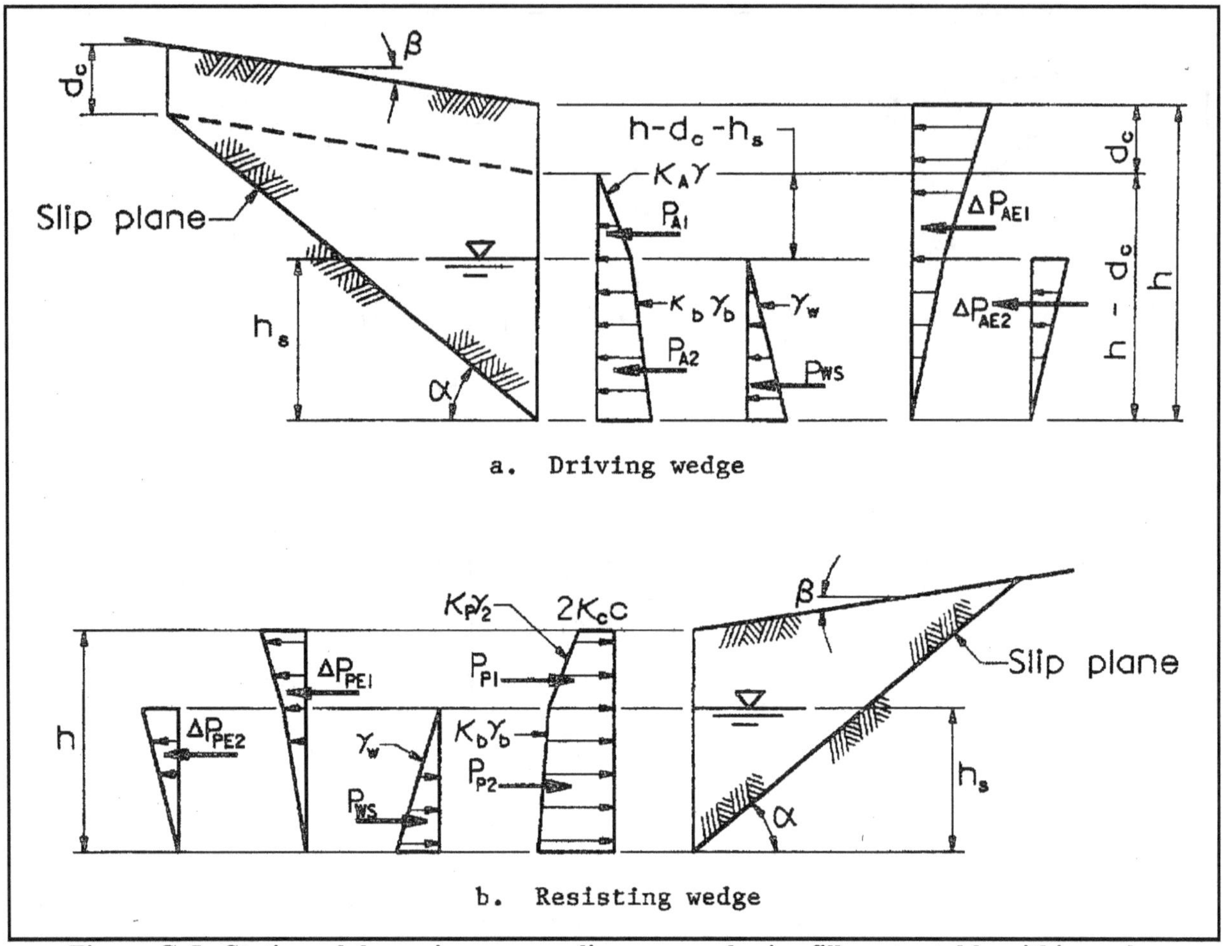

Figure G-5. Static and dynamic pressure diagrams, cohesive fill, water table within wedge

$$c_1 = \frac{2\tan\phi\,(\tan\phi - k_h) + \dfrac{4c\,(\tan\phi + \tan\beta)}{\gamma(h + d_c)}}{A} \qquad \text{(G-43)}$$

$$c_2 = \frac{\tan\phi\,(1 - \tan\phi\,\tan\beta) - (\tan\beta + k_h) + \dfrac{2c\,(1 - \tan\phi\,\tan\beta)}{\gamma(h + d_c)}}{A} \qquad \text{(G-44)}$$

$$A = (1 + k_h\,\tan\phi)\tan\phi + \frac{2c\,(1 - \tan\phi\,\tan\beta)}{\gamma(h + d_c)} \qquad \text{(G-45)}$$

$$d_c = \frac{c/\gamma}{\cos\alpha\,(\sin\alpha - \tan\phi\,\cos\alpha)} \qquad \text{(G-46)}$$

(2) Resisting Force. The static components for the resisting wedge are (Figure G-5b):

$$P_P = P_{P1} + P_{P2} = \frac{1}{2} K_P \gamma (h - h_s)^2 + \frac{1}{2} h_s \left[2K_P \gamma (h - h_s) + K_b \gamma_b h_s \right] + 2K_c ch \qquad \text{(G-47)}$$

$$P_{ws} = \frac{1}{2} \gamma_w h_s^2 \qquad \text{(G-48)}$$

and the dynamic components are (see Figure G-5b):

$$\Delta P_{PE} = \Delta P_{PE1} + \Delta P_{PE2} = k_h \left[\frac{\gamma h^2}{2 (\tan \alpha - \tan \beta)} \right] + k_h \left[\frac{(\gamma_s - \gamma) h_s^2}{2 \tan \alpha} \right] \qquad \text{(G-49)}$$

giving a total force of:

$$P_{PE} = P_P + P_{ws} + \Delta P_{PE} \qquad \text{(G-50)}$$

where γ, γ_b, γ_s, and γ_w are defined in paragraph 3-26c(4)(a), and

$$K_P = \left(\frac{1 + \tan \phi \cot \alpha}{1 - \tan \phi \tan \alpha} \right) \left(\frac{\tan \alpha}{\tan \alpha - \tan \beta} \right) \qquad \text{(G-51)}$$

$$K_b = \left(\frac{1 + \tan \phi \cot \alpha}{1 - \tan \phi \tan \alpha} \right) \left[1 + \left(\frac{\tan \alpha}{\tan \alpha - \tan \beta} - 1 \right) \frac{\gamma}{\gamma_b} \right] \qquad \text{(G-52)}$$

$$\alpha = \tan^{-1} \left(\frac{-c_1 + \sqrt{c_1^2 + 4c_2}}{2} \right) \qquad \text{(G-53)}$$

$$c_1 = \frac{2 \tan \phi (\tan \phi - k_h) + \dfrac{4c (\tan \phi - \tan \beta)}{\gamma h}}{A} \qquad \text{(G-54)}$$

$$c_2 = \frac{\tan \phi (1 + \tan \phi \tan \beta) + (\tan \beta - k_h) + \dfrac{2c (1 + \tan \phi \tan \beta)}{\gamma h}}{A} \qquad \text{(G-55)}$$

$$A = (1 + k_h \tan \phi) \tan \phi + \frac{2c (1 + \tan \phi \tan \beta)}{\gamma h} \qquad \text{(G-56)}$$

$$K_c = \frac{1}{2 \sin \alpha \cos \alpha \,(1 - \tan \phi \cos \alpha)} \cdot \frac{\tan \alpha}{\tan \alpha - \tan \beta} \qquad \text{(G-57)}$$

G-4. Inertia Force of Structure

The inertia force of the structure, including that portion of the backfill above the heel or toe of the structure and any water within the backfill which is not included as part of the backfill wedge, is computed by multiplying the selected acceleration coefficient by the weight of the structure and backfill. This force is obtained by multiplying the mass by the acceleration coefficient.

G-4. Selection of Acceleration Coefficients

a. Minimum Acceleration Coefficients. Preliminary estimates of horizontal acceleration coefficient values are listed in Table G-1. The seismic zone designations used in this table are defined in ER 1110-2-1806. These values can be used to determine if the lateral earthquake forces control the stability of structures. The vertical acceleration coefficient should be estimated as two-thirds of the horizontal acceleration coefficient. If failure of the structure would jeopardize the safety of a dam, then the acceleration coefficients should be consistent with those used for the stability analyses and concrete design of the dam.

Table G-1 Minimum Seismic Horizontal Acceleration Coefficients

Zone	Coefficient
0	0.00
1	0.05
2A, 2B	0.10
3	0.15
4	0.20

b. Acceleration Coefficients Greater than 0.2. If the design acceleration coefficient exceeds 0.2, a wedge method of seismic analysis may be excessively conservative, and a permanent displacement or a dynamic soil-structure interaction analysis should be performed. A method for computing the magnitude of relative structure displacement during a specified earthquake is described by Whitman and Liao (1985). The dynamic soil pressures and associated forces in the backfill may be analyzed as an elastic response using Wood's method as described in Ebeling and Morrison (1992).

G-5. Example

a. Problem definition.

Soil properties (on both sides of structure):

$\gamma = 0.12$ k/ft^3 (moist weight)
$\gamma_b = 0.0625$ k/ft^3 (buoyancy weight)
$\gamma_s = 0.125$ k/ft^3 (saturated weight)
$\phi = 35°$, $c = 0$

Seismic coefficients:

$k_H = 0.20$
$k_v = 0$

b. Find forces acting on driving side.

$$c_1 = \frac{2(\tan\phi - k_h)}{1 + k_h \tan\phi} = \frac{2(0.700208 - 0.2)}{1 + 0.2(0.700208)} = 0.877526 \qquad \text{(G-)}$$

$$c_2 = \frac{\tan\phi\,(1 - \tan\phi \tan\beta) - (\tan\beta + k_h)}{\tan\phi\,(1 + k_h \tan\phi)} \qquad \text{(G-)}$$

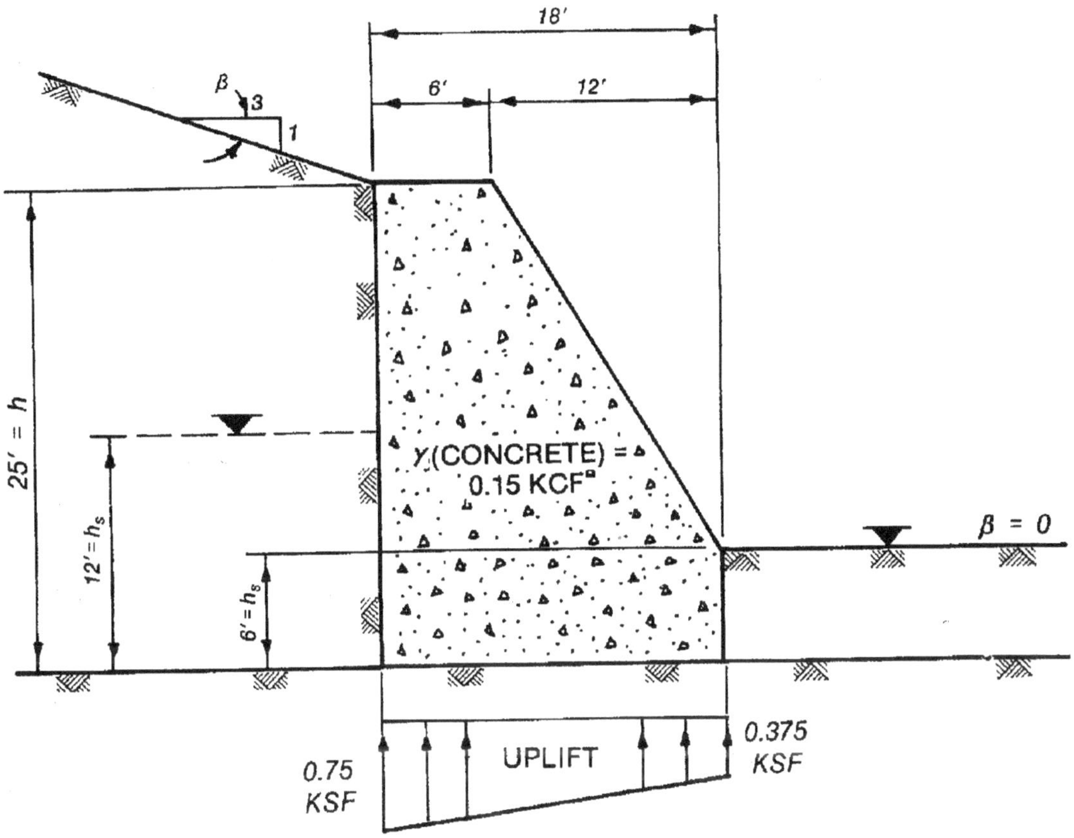

$$c_2 = \frac{0.700208\left(1 - 0.700208 \times \frac{1}{3}\right) - \left(\frac{1}{3} + 0.2\right)}{0.700208(1 + 0.2 \times 0.700208)} = 0.004315 \qquad \text{(G-)}$$

$$\alpha = \tan^{-1}\left(\frac{c_1 + \sqrt{c_1^2 + 4c_2}}{2}\right) = \underline{\underline{41.426^\circ}} \qquad \text{(G-)}$$

$$K = \frac{1 - \tan\phi \cot\alpha}{1 + \tan\phi \tan\alpha)} = \frac{1 - 0.700208(1.133240)}{1 + 0.700208(0.882425)}$$

$$K = 0.12763$$

$$K_A = K\left(\frac{\tan\alpha}{\tan\alpha - \tan\beta}\right) = 0.12763\left(\frac{0.882425}{0.882425 - \frac{1}{3}}\right) = \underline{\underline{0.2051}} \qquad \text{(G-)}$$

$$K_b = K\left[1 + \left(\frac{\tan\alpha}{\tan\alpha - \tan\beta}\right) - 1\left(\frac{\gamma}{\gamma_b}\right)\right] \quad \text{(see Appendix H)}$$

$$K_b = 0.12763\left[1 + \left(\frac{0.882425}{0.549092} - 1\right)\left(\frac{0.12}{0.0625}\right)\right] = \underline{\underline{0.2764}}$$

$$P_A = \frac{1}{2} K_A \gamma (h - h_s)^2 + \frac{1}{2}(h_s)\left[2K\gamma(h - h_s) + K_b \gamma_b h_s\right] \qquad \text{(G-)}$$

$$P_A = \frac{1}{2}(0.2051)(0.12)(13)^2 + \frac{1}{2}(12)\left[2(0.2051)(0.12)(13) + 0.2764(0.0625)(12)\right]$$

$$P_A = \underline{\underline{7.16\ k}}$$

$$\Delta P_{AE} = k_h\left[\frac{\gamma h^2}{2(\tan\alpha - \tan\beta)} + \frac{(\gamma_s - \gamma) h_s^2}{2\tan\alpha}\right] \qquad \text{(G-)}$$

$$\Delta P_{AE} = 0.2\left[\frac{0.12(25)^2}{2(0.549092)} + \frac{0.005(12)^2}{2(0.882425)}\right] = \underline{\underline{13.74\ k}}$$

$$P_{ws} = \frac{1}{2}\gamma_w h_s^2 = \frac{1}{2}(0.0625)(12)^2 = \underline{\underline{4.50\ k}} \qquad \text{(G-)}$$

c. Find forces acting on resisting side.

$$c_1 = \frac{2(\tan\phi - k_h)}{1 + k_h \tan\phi} = \frac{2(0.700208 - 0.2)}{1 + 0.2(0.700208)} = 0.877526 \qquad \text{(G-)}$$

From equation G-18

$$c_2 = \frac{\tan\phi - k_h}{\tan\phi\,(1 + k_h \tan\phi)}$$

$$c_2 = \frac{0.700208 - 0.2}{0.700208(1 + 0.2 \times 0.700208)} = \underline{\underline{0.626618}}$$

$$\alpha = \tan^{-1}\left(\frac{-c_1 + \sqrt{c_1^2 + 4c_2}}{2}\right) = \underline{\underline{24.999°}} \qquad \text{(G-)}$$

From Equation G-34

$$K_P = \frac{1 + \tan\phi \cot\alpha}{1 - \tan\phi \tan\alpha)} = \frac{1 + 0.700208(2.144605)}{1 - 0.700208(0.466286)}$$

$$K_P = \underline{\underline{3.7144}}$$

From Equation G-30

$$P_P = \frac{1}{2} K_P \gamma_b h^2 = \frac{1}{2}(3.7144)(0.0625)(6)^2 = \underline{\underline{4.18\,k}}$$

From Equation G-32

$$\Delta P_{PE} = k_h \left(\frac{\gamma_s h^2}{2\tan\alpha}\right) = 0.2\left[\frac{0.125(6)^2}{2(0.466286)}\right] = \underline{\underline{0.97\,k}}$$

$$P_{ws} = \frac{1}{2}\gamma_w h_s^2 = \frac{1}{2}(0.0625)(6)^2 = \underline{\underline{1.13\,k}} \qquad \text{(G-)}$$

d. Find inertia force due to weight of structure.

$$18' \times 25' \times 0.15 = 67.50 \times 12.50' = 843.75$$

$$-\frac{1}{2} \times 12' \times 19' \times 0.15 = \underline{-17.10} \times 18.67' = \underline{-319.25}$$

$$W = 50.40\,k \qquad 524.50$$

$$\bar{y} = \frac{524.50}{50.40} = 10.41\ ft$$

$$k_h W = 0.2(50.40\ k) = 10.08\ k$$

e. *Summary of forces and pressure distributions.*

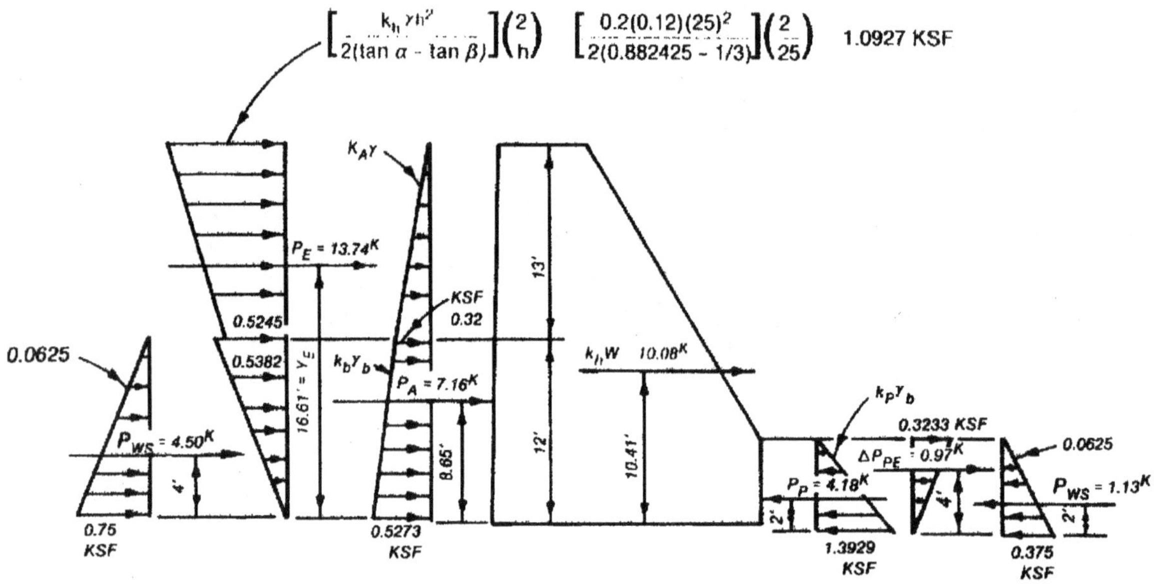

(1) Permissible simplification for dynamic earth pressure distribution—driving side. The discontinuity of this pressure diagram, at the water table, may be eliminated by considering that the soil weight above and below water is equal to the moist weight. The difference is not significant. In this case, the difference in forces is -0.58% and difference in dimension, Y_E, is +0.36%.

(2) Mononobe-okabe force and pressure distribution--resisting side. If the pressure diagrams for P_P and ΔP_{PE} (on the preceding page) are combined, negative pressure will be obtained for some distance below the top of ground. Since earth pressure cannot pull on the structure, the pressure diagram and force should be determined by setting all negative pressures to zero.

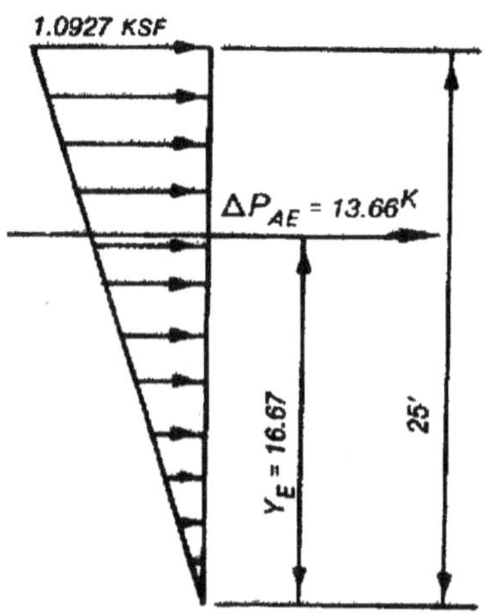

Appendix H
Classification of Structures

H-1. General

Critical structures, per ER 1110-2-1806, are those which are part of a high hazard project and whose failure will result in loss of life. Loss of life can result directly due to flooding, or indirectly from secondary effects. Structure classification is to be based on a total project evaluation which considers all project features, their interdependence, and the impact substandard performance of one project feature might have on the importance and criticality of other project features. A critical structure determination involves consideration of the possibility of failure, and the potential for loss of life should failure occur. Under certain circumstances the population in the vicinity of the failed structure may not be at risk, or if at risk, there may be sufficient warning time to evacuate the people from downstream areas that will be inundated. Various earthquake, flood, and latent deficiency failure condition scenarios must be examined to determine if failure can result in loss of life. Critical structures are subject to more stringent sliding safety factor requirements. The types of hydraulic structures, which could be classified as critical, include gravity dams and spillways, arch dams, urban flood walls, coastal flood walls, and intake towers. Structures not qualifying as critical structures shall be classified as normal.

H-2. Classification Process for Critical Structures

a. Project characteristics. Various project characteristics should be investigated to determine how they may affect project performance during major flood and earthquake events. The characteristic of the impoundment area and of the downstream reaches of the project are especially important to safety. For flood control projects, it is likely that conditions conducive to landslides, subsidence, and erosion will be most critical during extreme flood events when the impoundment is at its highest stage and project discharge is at a maximum. Site geology and seismicity are important when evaluating earthquake ground motions, sliding stability, liquefaction, and earthquake induced fault displacements. The quantity of water impounded by the project, distance to populated areas, and the number of people at risk are important when evaluating life safety performance. Project characteristics to be considered during the safety evaluation include:

b. Failure scenarios. All potential structure failure scenarios must be investigated to determine their impact on structure performance. Erosion and piping can adversely affect structure stability and structure performance. When structures are located in areas of high seismicity, earthquake effects must be considered in the failure scenario investigation. Earthquake effects include ground motion demands, fault displacements, subsidence, slope instability, and liquefaction.

c. Failure consequences. Structures, which fail to meet performance objectives, can result in loss of life, or damage to property, essential lifelines or the environment. Loss of life is the only consequence applicable to the structure classification process. In general, project performance objectives are to:

- Retain and release impoundments in a planned regulated manner.
- Prevent structure damage under usual and unusual load conditions.
- Prevent structure collapse under extreme load conditions.
- Allow adequate time under emergency conditions to evacuate people from areas subject to flooding.
- Remain operational to permit a controlled release of impounded water following major flood and earthquake events.

d. Structure classification. The consequences of potential failure will determine whether a structure is to be designated as *critical*, or *normal*. The *critical* structure designation is only to be used in those cases where failure of the structure to perform will directly or indirectly lead to loss of life. Where the lives of people are not at risk, or when there is time to evacuate people from locations where they would be a risk should failure occur, the structure shall be classified as *normal*.

H-3 Examples of Classifications.

a. Gravity dam - flood control project.

(1) Project characteristics. The project consists of a straight axis gravity dam and ungated overflow spillway. The dam has no permanent reservoir. The purpose of the dam is to protect against flooding resulting from winter storms and thunderstorm runoff. The dam is designed to detain flood flows, and to release stored water in a regulated manner so as not to cause property damage and loss of life. The dam has the capacity to store 1,500,000 cubic meters (1,260 acre feet) of water. A sudden release of stored water could result in loss of life. The canyon upstream of the dam has steep side slopes and sparse vegetation. The dam reservoir can be filled to capacity in a matter of one or two hours following a major thunderstorm. The canyon downstream of the dam is also steep and narrow and extends to the edge of populated areas. The dam is located two miles upstream of a town that has a population of 23,000. Approximately 5,000 people live in the area that would be inundated should the dam fail. It can be expected that most, if not all, of those people would be at risk.

(2) Failure scenarios.

(a) Latent deficiencies. Since there is no opportunity to fill the reservoir under controlled conditions to monitor displacements and uplift pressures, the dam is considered potentially susceptible to failures caused by unknown latent deficiencies, deficiencies that may not be discovered until flooding occurs. The warning time will be short since the reservoir will fill rapidly under thunderstorm runoff conditions.

(b) Earthquakes. The dam is located in an area where major earthquakes are possible. However, the dam is dry 95-percent of the time so that dam failure and loss of life due to an earthquake related failure is highly improbable.

(c) Floods. The potential for a flood type failure exists for extreme conditions where the water overtops the dam and the discharge is of sufficient magnitude and duration such that erosion of dam foundation material occurs. Warning times for this type of failure should be sufficiently long to evacuate people from the downstream area that would be inundated. Failures due to excessive uplift and piping are possible failure scenarios for which warning times would be short and not sufficient to evacuate people from the downstream area that would be inundated. These type of failure mechanisms, although extremely unlikely, should be considered since there will be no opportunity to evaluate dam performance under controlled pool raise conditions.

(3) Structure classification. Monitoring the reservoir and performance of the structure under controlled conditions is not an option. Sudden failure and release of impounded water, although extremely unlikely, is possible. Warning times under such circumstances will not be sufficient to evacuate people from the downstream areas that would be inundated. The dam is considered to be a *critical* structure, and therefore the safety provisions applicable to *critical* structures must be used for all flood loading conditions. The safety provisions applicable to *normal* structures may be used for earthquake loading conditions since failure due to earthquake ground motions will not lead to loss of life.

b. Gravity dam - hydropower project.

(1) Project characteristics. The project consists of gravity dam monoliths (non-overflow sections), generator bay monoliths, and gated spillway monoliths. The project routinely impounds water to within 2 feet of the top of spillway gates to maximize power benefits. Flood storage benefits are small because the capacity to store flood flows is limited and because there are few people living in areas subject to flooding. The project has the capacity to store 0.50 cubic kilometers (400,000 acre-feet) of water. Water routinely impounded for hydropower is equal to 0.43 cubic kilometers (350,000 acre-feet). When flooding is forecast the reservoir is lowered. Storage up to 0.12 cubic kilometers (100,000 acre-feet) can be provided for flood protection. The canyon upstream and downstream of the dam has steep side slopes and sparse vegetation. All populated areas in the vicinity of the project are located above the canyon rim and therefore not subjected to flooding. The downstream canyon widens out about 45

kilometers (28 miles) from the dam. The nearest town that would be prone to flooding is located 50 kilometers (31 miles) downstream of the project. Should the dam fail there would be approximately 10 hours to notify and evacuate people from the area that will be inundated. Approximately 100,000 people live in the town located 50 kilometers (31 miles) downstream of the project. . It can be expected that approximately 2000 of the people would be at risk due to flood inundation.

(2) Failure scenarios.

(a) Latent deficiencies. The reservoir is kept within two meters (6.5 feet) of the top of dam. The hydrostatic loads that are routinely applied to the dam are 95-percent of those estimated for extreme flood events. Therefore the possibility that there are unknown latent deficiencies that could lead to erosion, slope instability, subsidence, piping, and other conditions that might impair safety is extremely remote.

(b) Earthquakes. The dam is located in an area of low seismic activity. Earthquake loadings will not control the design.

(c) Floods. The potential for a flood type failure is extremely unlikely. The spillway is sufficient to pass the probable maximum flood (PMF). The stilling basin is designed to accommodate heavy discharge conditions and is inspected on a regular basis. Failures due to excessive uplift and piping are unlikely failure scenarios since the project is routinely subjected to high heads, and since periodic inspections and monitoring of project instrumentation will catch excessive uplift and piping conditions well before such conditions can lead to failure.

(3) Structure classification. Regardless of failure scenario, warning times would be sufficient to evacuate people from downstream areas that would be inundated. The dam is considered to be a *normal* structure and all safety requirements will be those applicable to *normal* structures.

c. Urban flood wall.

(1) Project characteristics. The project consists a 4-meter (13 foot) high concrete I-wall that is part of a 10-km (6-mile) levee system providing flood protection to an urban area. The project is located adjacent to a major river, which has exceeded flood stage on many occasions. The wall and river bank are subject to erosion during flood stage since neither have riprap protection. The project has the capacity to prevent 600,000 cubic meters (500 acre-feet) of water from inundating populated areas. The unprotected side of the levee-flood wall project consists of a river basin and agricultural farmland. The protected side is a densely populated urban area. The urban area is immediately adjacent to the levee-flood wall project. Approximately 20,000 people reside in the area that would be inundated if the flood wall should fail.

(2) Failure scenarios.

(a) Latent deficiencies. Erosion of the river bank and levee can occur during flood events. This erosion can lead to failure of the wall during intermediate river stages (when overtopping of the wall will not occur).

(b) Earthquakes. Earthquake failure is unlikely. If the wall failed due to earthquake ground motions there would be no loss of life.

(c) Floods. There is a potential for the wall and levee to be overtopped during an extreme flood event. However, under these conditions there is sufficient time to warn and evacuate people residing in flood prone areas. Overtopping can lead to local wall and levee failures.

(3) Structure classification. The flood wall is considered a *critical* structure since the wall could fail suddenly without warning during intermediate flood stage conditions. Therefore the safety provisions applicable to *critical* structures must be used for flood loading conditions. Since failure due to earthquake ground motions will not lead to loss of life, the safety provisions applicable to *normal* structures may be used for the earthquake loading conditions.

d. Intake tower.

(1) Project characteristics. The intake tower is part of a flood control project that consists of a 85-meter high earth-and-rockfill embankment dam, a side channel regulated spillway, outlet works, and intake tower access bridge. The intake tower is founded on rock, is 80-meters (260 feet) high, and rectangular in shape. Flow regulation through the intake tower - outlet works is accomplished with two slide gates located at the upstream end of the outlet works tunnel. Although used primarily for flood control, the pool for most of the year is maintained at a level that submerges 40-meters (130 feet) of the tower. The pool of record resulted in 60-meters of tower submergence. Maintenance bulkheads are provided in the intake tower to allow the tunnel, outlet works structures, and slide gates to be inspected and maintained. The maintenance bulkheads must be placed under balanced head. The outlet works tunnel is 5 meters (16 feet) in diameter and located in the rock abutment for the dam. The time required to drawdown the pool, in case it should be necessary to make repairs to the embankment dam, is very long (approximately 2-months). Under maximum pool conditions the dam impounds 0.11 cubic kilometers (90,000 acre-feet). Under recreational pool conditions the dam impounds 0.07 cubic kilometers (60,000 acre-feet). The intake tower is surrounded by water and therefore always under balanced head conditions. Both the upstream and downstream river channels are steep and heavily wooded. Urban populated areas exist within 10 km (6 miles) downstream of the dam. If water were suddenly released from the project the flood wave would reach populated areas in 45 minutes. This not considered to be adequate time to evacuate people even under ideal conditions. Should dam failure occur, the lives if 100,000 people would be at risk.

(2) Failure scenarios.

(a) Latent deficiencies.. The intake tower is surrounded by water and therefore always under balanced head conditions. The potential for latent deficiency failure mechanisms developing during flood events is extremely small. It is highly unlikely that during major or extreme flood events debris from landslides in the upper basin would block tower intakes and reduce discharge capability. The spillway has the capacity to pass PMF flows without the additional discharge capacity provided by the outlet works. Overtopping of the dam is not possible, even under conditions where the tower intakes are blocked by debris.

(b) Earthquakes. The project is located in a high seismic area. There is a 50% chance that severe damage, and possibly collapse of the tower and access bridge could occur during a major earthquake.

- *Tower collapse.* Since the tower is operated from a remote location there will be no direct loss of life due to tower collapse, nor will collapse of the tower cause loss of life due to a sudden release of pool. Should tower collapse and the outlet works tunnel exposed to ungated operation, the flow released downstream under open channel flow conditions will be insufficient to cause erosion that could lead to dam failure.

- *Dam failure due to impaired tower drawdown capability.* Earthquake ground motions resulting from major or extreme earthquake events will weaken the dam resulting in subsequent failure due to seepage and piping of impervious core material. Pool drawdown through the outlet works (even when the outlet works is discharging at full capacity) will not occur at a rate sufficient to prevent dam failure

- *Dam failure due to outlet tunnel failure and the inability to make an upstream closure.* Under certain damage scenarios it is possible that water escaping the outlet works tunnel could lead to embankment dam erosion and failure. This could occur if the tower were damaged severely enough to prevent an upstream closure, and if the outlet works tunnel, due to fault displacements was damaged severely enough to allow water to escape onto the dam embankment. Since the outlet works tunnel is located in a rock abutment and since potential water transmission paths will not lead to embankment dam erosion, this failure scenario is extremely unlikely.

(c) Floods. The intake tower is surrounded by water and therefore always under balanced head conditions. The potential for tower failure during extreme flood events is extremely small.

(3) Structure classification. The various failure scenarios investigated suggest that loss of life due to intake tower failure is extremely unlikely, and that the tower is not critical with respect to overall project performance. The intake tower is therefore classified as a *normal* structure.

e. Navigation lock - upstream gate monolith.

(1) Project description. The project consists of a 33.5 m (110 foot) wide by 206 m (675 feet) long navigation lock with the capability to accommodate an 25 m (82 feet) lift. The lock is founded on a shale formation, which is known to have clay seams with low shear strength. The reservoir upstream of the lock and dam impounds up to 0.62 cubic kilometers (500,000 acre) feet of water. The basins upstream and downstream of the project consist of steep hillsides with marginally stable slopes. Landslides have occurred when hillsides are saturated by rainfall. The navigation lock is located in an urban setting. The hillsides upstream and downstream of the project have been developed for residential and commercial use. Subsidence and landslides are possible under rapid drawdown conditions that could result from a sudden loss of pool. Approximately 15,000 people are at risk.

(2) Failure scenarios.

(a) Latent deficiencies. Unexpected movements along a deep seated clay seam layer could lead to failure of the gate monolith and loss of pool. This could occur during normal operation with the lock chamber at tailwater (maximum differential head during normal operation). Failure could also occur when the lock chamber is unwatered for maintenance (maximum differential head condition). Failure could be sudden without warning, and without time to fill the lock chamber to equalize the differential head.

(b) Earthquakes. The project is located in a low seismic area. Failure of the gate monolith due to earthquake ground motion is extremely unlikely.

(c) Floods. The project spillway has the capacity to pass the inflow of major floods without overtopping the project structures. Flood induced failure is extremely unlikely.

(3) Structure classification. Since there is a high probability that failure of the gate monolith will indirectly cause loss of life, the gate monolith is considered a *critical* structure. Sliding factor of safety requirements for *critical* structures therefore will be used for *normal operating*, and *maintenance* load conditions.

www.ingramcontent.com/pod-product-compliance
Lightning Source LLC
LaVergne TN
LVHW080527210926
848481LV00022B/3406